Cici

To all the struggles and how we have overcome!!
Against All Odds!
Love
Tanikka Price

Against All Odds:
From Teen Mom to Boss Lady

For more information, email info@melaningracepub.com

ISBN 9781736987988

Hey! Get our free Boss Mom Affirmation Worksheet!

This will help you speak positive words over your life and your children!

https://bit.ly/bossmomaffirmations

FREE BOSS MOM AFFIRAMTIONS

Daily affirmations for kids and adults to help you and your family change your life for the better. Positive thinking is the foundation of success and happiness, so start speaking positive now!

BIT.LY/BOSSMOMAFFIRMATIONS

DEDICATION

To all the young women who've started the journey of motherhood, we empathize.

All hope is not lost! Continue to focus on your goals.

You can achieve anything you want.

TABLE OF CONTENTS

FOREWORD

"She remembered who she was, and the game changed."
-Lalah Delia

Leaders, servants, innovators, strategic planners who execute with precision, creative designers, transformative authors, authentic speakers, and BOSSES are just a few words that describe the women in this book. While these words are connected directly to them now, it wasn't always this way. The strong women you are going to read about began their journeys as teenage mothers. Some had low self-esteem and struggled against being abandoned and ignored by family & friends. Others had to deal with the rejection of their child's father and were forced to raise a baby all on their own. The beauty of these stories will give the ugly truth of the journey to becoming high-profile attorneys, entrepreneurs, doctors, global brand leaders, and corporate officers.

I will never forget the day my parents told my elite high school principal that I was pregnant. I was transferred to an alternative school that same afternoon. My spot in the magnet program was gone, and it devastated me. This alternative school was nothing like the place I loved as a freshman. There were daily fights, no windows, always disgusting girls' bathrooms, and let's not mention the classwork. I was back in a position where I was no longer being challenged. I cried every single day. I begged God to kill me so I wouldn't have to go through being bullied by the other girls because I came from a "fancy" school. The bullies were my fuel to never step foot in that school again. I was determined to get back on track, get my spot back at my high school, and win big at life. I was not going to be defeated. Sometimes we create our own storms, but it doesn't mean we can't move the clouds when we are ready for the sun again.

Even though this was the hardest experience I have ever faced, the desire to be successful was forever present. As a member of a sales leadership team and owner of a wellness brand, the stories of *Against All Odds* pierced my heart. When you sit down to read this book, you'll want to grab your favorite beverage and a comfortable spot. You are going to want to read all the stories at once because it's just that powerful. Be ready to be encouraged and inspired by stories that bring wisdom to anyone experiencing struggle and need a light to make it to the end of the tunnel.

Dr. Yolanda Y. Cox

Transformation Catalyst

Let it Out Brands

"We help women in leadership, and corporate-prenuers get off the weight-loss gimmick train and S.H.I.F.T to a healthy lifestyle that fits their real-life through our successful system that focuses on mindset, heart matters, health goals, and soul wealth."

INTRODUCTION

A generation ago, unwed teenage pregnancy was uncommon. Pregnant teenage girls were shunned and sent off to live with family members in remote parts of the country to prevent shame from befalling their im-mediate families. Teenage pregnancy has become more pervasive in the wake of changing societal norms and popular teen pregnancy shows. Teenage pregnancy has become the rule versus the exception in some communities, and it is surprising to see a young lady graduate high school before becoming a mother. It is safe to say that most of us know someone who became a teenage mother. The truth is, teenage pregnancy can happen to anyone who is fertile and sexually active. According to the Center for Disease Control, although teen birth rates have declined, U.S. teen pregnancy is still significantly higher than in any other country. There are many disadvantages to teen pregnancy including, but not limited to:

- higher high school dropout rates
- increased physical and mental health issues
- higher unemployment rates
- increased likelihood of incarceration
- increased occurrences of poverty
- higher risk for substance abuse

8.25% of teen moms have a second child within 24 months, significantly increasing their riskiness of realizing one or more of these disadvantages (CDC.com). Unfortunately, these rates are drastically higher in minority communities.

Against All Odds is not the typical educational book about teenage pregnancy. It does not hide the experiences of teenage pregnancy, nor does it glorify them. Instead, its goal is to help teenage and young mothers navigate their own experiences by traveling through the journeys of 10 women who didn't allow teenage pregnancy to destroy their lives nor define them. These stories articulate the struggles, misguidance, and repercussions of being a teenage mother and the ultimate transformation into a boss lady.

A boss lady is a woman who sets her own rules, dominates in her respective field, and lives her best life. No, she is not perfect, but she has learned to love herself and play big in her purpose. These authors' stories walk you through EXACTLY how they transformed their lives and overcame the negative stigmas that many teenage mothers suffer. You will laugh, cry, and experience actual change throughout these stories. Our greatest request is that you apply what you learn.

> "Knowledge without action, is like having no knowledge at all."
>
> -Ted Nicholas

If you are a teen or young mother reading the book, we hope that you're able to identify yourself in these stories. Underneath being a teen mom is also the journey of overcoming adversity, low self-esteem, identity struggles, neglect, abandonment, rejection, physical and emotional abuse, loneliness, and failure. In most stories, being a teen mom was secondary to these other issues. It doesn't mean if you become pregnant at a young age that you suffer from any of these issues; we want to be sure we equip you with the right tools to transition into a boss lady.

Through our experiences, here is our key takeaway and what we want you to get out of reading this book:

Realize your life isn't over being a young mother.

Understand that you have the power to live your dreams.

Persevere through unforeseen circumstances.

Prioritize your life.

Don't settle for less.

and most importantly, BEAT THE ODDS!

FROM LOVING HIM TO FINDING ME

Dr. Janell Jones, MSW, LSW

Underprivileged

I grew up in a low-income project in Columbus, Ohio. I was the youngest of six siblings. My upbringing was not perfect as we struggled in poverty and lived in a drug-infested community. I had suffered a significant loss by the time I was five years old. The people closest to me, my father and my sister, were murdered because of domestic violence. To make matters worse, many of my family members were addicted to some sort of substance. I fell into the statistics of being a teenage mother at 16. Yet, my story doesn't end there. I graduated from The Ohio State University with bachelor's and master's degrees. I became a certified life coach, licensed clinical therapist, international bestselling author, course creator, and a serial entrepreneur.

The Boy Next Door

Let's go to the beginning of my pregnancy story. They considered me the girl that no one ever thought would get pregnant as a teenager. I loved going to school and did things pretty much by the book until I started dating at 14. I had moved out of the low-income projects where I had lived most of my life, into another not-so-great neighborhood. I started missing the people in my old hood, so I would frequently visit. If I didn't visit, I would call to catch up. There was one guy, let's call him boy, that all the girls in the neighborhood wanted. He was tall, brown-skinned, had light hazel eyes, and had "good hair." He was very charming and knew how to "holla" at girls. That didn't impress me much. I thought he was immature, and I didn't want to be bothered. He would always approach me. If there was a group of us girls standing in a circle, he would come to me first and try to hug me. Of course, the girls didn't like that, but in my mind, it just wasn't a turn on. He was just a friend like most boys in the hood. We grew up together and had known each other since we were about 5 or 6 years old.

In my circulation of phone calls, I decided to call him. I wanted to surprise him when I called. His grandmother had the same phone number for literally 30 years, and it was easy to remember. The phone rang, and his dad picked up. I said hello and asked if I could speak to that boy. The boy got on the phone, and I said hi, and he said hello back and asked who I was. I responded with who do you want it to be and he said, Janet Jackson. We laughed, and then he said I know who you are; I recognize your voice. As we were getting off the phone, he asked me to be his girlfriend. To be honest, I didn't know what to expect. I never saw myself with him, but somehow "yes" slipped out my mouth. Initially, I thought this was a joke, and we'd break up within a few months, but things started getting serious.

We started moving very fast and spent the next few months ignorantly in love and became wrapped up in each other's world. He was my everything and I decided to give him my virginity. He assumed the role of my father; he became my protector. I was starving for love and didn't realize it until I got older. He quickly filled in the gap. No one could tell us anything. Many people would say it was 'puppy love.' But I would venture to say it's what we knew as love. He was my supporter. In fact, he would challenge me to do better in school. In the 10th grade, I achieved honor roll.

When I was 16 years old (he was 17), we recognized we needed to practice safe sex and get on some form of contraceptive. We didn't want to be teen parents and on my side of the family most women were. We caught the bus to Planned Parenthood where we got condoms, and I started taking birth control. My mom did not know that I was sexually active so I tried to remain discreet. I was excited to learn that I didn't have to be 18 or older to get birth control from Planned Parenthood. I could keep my little secret of not being a virgin.

After spending almost two years together, things fell apart. I can't pinpoint precisely what happened, but we started drifting apart and decided to amicably split. We would talk occasionally, but nothing serious. The time apart was hurtful since our world revolved around each other. After a few months of being broken up, I stopped taking birth control pills. I thought that since we were not having sex, there was no point in continuing to take them; I wasn't having sex with anyone. One day, he came over before I went to work. I am still not sure what the point of the visit was. We had sex one last time. We were so caught up in the moment that we didn't think about any protection.

About a month passed, and my mom was asking me questions about my period. I noticed little, but according to her, I was snappier. Her saying that concerned me, so I was secretly trying to confirm if I was pregnant. I took a home pregnancy test which came out positive. I was trying to avoid going to my family doctor to confirm the pregnancy because I knew he would be disappointed with me. I had been going to him since I was about eight yrs. old. To avoid seeing him, I walked into a clinic closer to my house, who denied me the same-day appointment because I wasn't an established patient. I had to go to my family physician, anyway. Well, you›re pregnant, he said, and your due date is June 18th. I told my ex about the pregnancy, and to my surprise, he denied that the baby was his. I couldn›t believe that the person I'd known most of my life and who was the only person that I had been with, was denying our baby. It hurt me to my core, especially since he chased me most of my life. I was afraid that I was going to have to face being a teen mom alone. As close as we once were, and all that we'd been through, he left me high and dry.

When we were dating, we made plans to go to the career center together during our junior year. By the time our junior year started, we had broken up; I was pregnant, and he had denied our baby. When I would see him in the hall, I'd feel so embarrassed and hurt. He would ignore me like I didn't exist. After getting the paperwork from the doctor with my due date, I made a copy to give to him. I was standing at my locker at the career center and asked his friend to give him the paper. Moments later, after class started, he and his friend came into the room and tossed the ripped pieces of the pregnancy confirmation on the floor. My heart sank, but I got out of my seat, dropped on my knees with my body shaking, and picked up each shredded piece of paper as I held back the tears in my eyes.

After confirming with my mom that I was pregnant, she told me I disappointed her, but she encouraged me to finish high school. She was also mad at Boy for denying our baby because of our extensive family history. His aunt was respectfully questioning my mom about how the pregnancy happened, being that we were to be supervised, but we were having sex everywhere. Every time we saw each other, we were sneaking to find a way to have sex.

He ultimately apologized and came around when I was five months pregnant. He started attending doctors' appointments with me. He admitted that he was afraid, which was why he was in denial. Even though we were still broken up, he did his best to be in this child's life. A lot happened in-between time. We went our separate ways with other people, but continued to co-parent. We eventually got back together, got married, and had two additional kids. By the time I was 22, I had three kids. This was a very trying time. It was extremely challenging to navigate through life with three kids and a marriage on the rocks. Being a young mother of three children-and not knowing who I was, was like driving to a destination with no GPS. I was lost and confused. Not only from what I went through with my ex, but the fact that my self-esteem and self-worth was shot.

The Hard Knocks Life

Sometimes I felt like giving up and quitting. Growing up in poverty was a disadvantage, and I didn't want my children to be subjected to the same. I didn't see how that was going to happen. Of course, I wanted more, but didn't have the direction to get there. What I saw growing up was the cycle of poverty. If anyone in my immediate family had dreams, they were not shared, and they definitely weren't implemented. The closest thing I remember was my oldest brother being a talented mechanic, but having no formal training. The word on the street was that he was so knowledgeable at fixing cars that he could work in the pit at professional race car events. However, he only got as far as fixing cars as a side hustle and the older he got, the less he did it. The ball was in my court to figure out how to escape the trap of poverty. I didn't think I could do it. I struggled in school with reading and writing. Although I enjoyed going to school, reading wasn't enjoyable. I remember having to go to the "special classes" to get additional resources for reading because I did so poorly on standardized testing. My reading and writing scores were high enough not to be placed in special education but low enough not to be in regular classes. As an adult, I realized I had an undiagnosed learning disability. I didn't feel well enough to produce a better life for my children even though I desired it.

No Longer A Victim

Unfortunately, as women, it's not as easy for us to turn our backs on our babies. When my ex threw those papers back into my classroom, I realized I was going to have to go on this journey alone. Yes, I had the support of my family, but this child was ultimately my responsibility. This changed my perspective. It made me realize people will disappoint you, and I had to pivot. I had to make a better decision for the sake of me and my babies. I went through an arduous divorce and became a single parent. I was unhappy with my situation and embarrassed that I couldn't offer my children more. I recall sitting on the floor in my apartment on the computer. I couldn't afford a computer desk, but I enrolled myself in college. I read Jack Canfield's book *Success Principles, How to Get From Where You Are to Where You Want to Be*, and there was a famous quote in there. "People change when the pain of remaining the same is greater than the pain of change."

That meant that people usually wait until they can't take it anymore to do something about their situation, and that's where I was. I didn't want to live a mediocre life, and I could not fathom looking my children in their eyes, requesting them to do more with their life if I hadn't with mine.

Making My Seat at the Table

With my background, statistics say I wasn't supposed to make it. However; God had bigger and better plans for me. I started to realize that I can have the life that I desire if I put in the work to make it happen. Here are my personal steps that transformed my life from teenage mother to boss lady.

1. ***Accept me.*** I had to accept myself to move forward with the fact that I was a teenage mother. Accepting myself allowed me to dig my heels into completing my goals and not accept the judgment of others. This is a critical step in anything that you do. It didn't mean that I was perfect. It meant that I accepted myself, flaws and all!

2. ***Strengthen my relationship with God.*** I don't mean to sound cliché here, but I had to go to the source of my creation. I had no clue who I was, and I became everything else; I lost myself. Even after my firstborn and having two additional children. I felt lost, which caused me to make decisions that had nothing to do with my purpose because I constantly sought approval from others.

3. ***Heal.*** I had to learn to heal from the pain of my past, whether it was self-inflicted or inflicted by others. Healing was a key principle with moving forward in my life. People look at healing as allowing oth-

ers to get away with wronging you. But it's really to benefit you. When we remain in a state of being unhealed, that leads to physical, mental, and emotional issues.

4. ***Dream Big***. I had to go places that I had mentally never been before. My upbringing only allowed me to think to a limited degree. Heck, I never went on vacation. My mother didn't drive, so we were always on the bus. I learned that I had to expose myself to more using visualization. I learned that meditation and visualization are powerful keys, and they got me from where I was to where I wanted to be. 4.

5. Do it afraid. There will be moments in your life where fear will arise. In fact, these moments happen often. Fear is a natural, human emotion. People tend to not move past that emotion, which paralyzes them from moving forward-living their dreams. When you push past fear by taking action towards your goals, you will increase your confidence, and taking action becomes easy. Then, you become unstoppable. Don't coward down to fear.

6. ***Pray over your children.*** My ex and I came from broken homes, which meant generational curses were passed down, even teenage pregnancy. I prayed over every aspect of my children's lives: purpose, friendships, money, business, and relationships. Although I escaped several generational curses, I didn't escape them all. Some still try to take hold of my children, especially in their thoughts.

It Stops Here!

Now, I can proudly say that I have broken generational curses, and it wasn't easy. Many years after graduating from high school and three kids later, I went to college and earned an associate degree in psychology. I earned bachelor's and master's degrees in Social Work and am a licensed clinical therapist in Ohio. I started my publishing company in 2017, and have written and published six books. I am a book coach, teaching others how to write and publish their books. I created and facilitated a youth program called Mahogany and Beautiful Bright Me to help girls with self-love, building confidence, goal-setting, and conflict resolution. I've created affirmation t-shirts and coffee mugs for parents with this program. I am a certified life coach, and I help women monetize their passions, skills, and expertise. I am a motivational speaker who spreads the healing message. I have a podcast entitled *Girl; You Got This*, where I spread empowering messages to people across the globe. * earned an honorary doctoral degree in leadership in healthcare, so now I am Dr. Jones. I shared my story honestly with my children, hoping they would learn from my mistakes. I held nothing back, and I was transparent about the part I played in my story, which was huge. I had to decide what I truly wanted out of life. It's easy to play the victim and "allow" things to happen to you without taking responsibility. I could have easily blamed my setbacks on not having a father, being raised in a drug-infested home with lack of affirmation, and poverty. Although those things affected my life, they didn't define me. I had to become clear on who I was. I got to know myself and my purpose.

As I became confident in myself, it permitted my children to do the same. My oldest son, whom I became pregnant with at 16, is 25 years old at the time of this writing. He currently doesn't have any children. He started his own publishing company and is dope with writing. He thrives in poetry and rap. My daughter started two businesses by the time she was 19. She is currently in the hair industry and is killing the game. She is passionate about beautifying others and bringing the best out of them. Her motto is excellent vibes only and to help other women thrive. She doesn't take part in drama and is focused on getting the bags. My youngest son is in barber school and is learning how to invest in real estate. He graduated high school during the pandemic and is now focusing on his life. He started his own business. My children didn't have it easy and went through a series of traumatic events, but they are doing things they love to do, and each has started businesses.

Purpose-Driven Life

Becoming a teenage mother was very difficult, I will not lie. I won't sit here and tell you that being a teen or young mother doesn't create its challenges. Being a teenage mother affected my life, but it didn't disable me. I had to learn the importance of prioritizing my time. If I were to do it all again, I wouldn't have wasted so much time trying to figure other people out. I was too busy trying to keep people in my life who didn't belong. I would have focused more on getting to know myself. When you have a child at a young age, you barely know who you are. Getting to know yourself isn't something that happens automatically; it's intentional. I know people who are well into their 40s and 50s who don't know themselves. Getting to know yourself creates a deliberate target. When we are unsure who we are, we aimlessly hit targets that have no purpose and that don't serve us.

I am grateful to share my story of overcoming teenage pregnancy. As I healed, I saw my blind spots and where I went wrong. Most of my choices stemmed from low self-esteem and the lack of love that I had for myself. Many factors were missing in my life, and having a baby didn't teach me how to love, especially since loving myself is a prerequisite to loving others. I didn't give up on myself, even in times where I counted myself out. Spend a great deal of time getting to know yourself. Count on *Momma and them*' or community resources to discover who you are and why you were created. Our greatest pleasures in life come when we fulfill our passions. Having a child doesn't mean you lose yourself. Your children do contribute to your Why! Your why is the reason you get up in the morning, you grind, you try, you push, you overcome, you conquer, or you succeed. Your why can be your children, parents, circumstances, or inner child. Your WHY is your life's biggest motivator.

Love, Boss Lady Dr. Janell Jones

FROM TEEN MOM TO TRAILBLAZER

Dr. Cortesha Cowan

Where it Started

Let me start by saying why I chose this title, "From Teen Mom to Trailblazer". Because in our society, teen mom carries such a stink. People think that because you are having a child at a young age that your life is over. Basically, you have already written the chapter in the book that says you will not be or do anything in your life. Well, I am here to take you on a brief journey into my life. Let us start with who is Dr. Cortesha Cowan.

Hello, I am Dr. Cortesha Cowan, Founder/CVO of Mothers Helping Mothers Inc. I was a teen mom who grew up in Lima, Ohio. I have leveraged my professional and personal experience to empower and help teen mothers, young girls, and women grow. My passion and drive for seeing teen mothers affected by homelessness pushed me to open the door and establish safe housing at the Haven of Hope crisis and transitional house. I have continued to do the work; my work now extends to Cape Town, South Africa, and Atlanta GA.

A Small-Town Girl

I grew up in a small town, Lima, Ohio. My family was close-knit but, like many families, dysfunctional. Back then, I grew up in the crack-cocaine era. Drugs and gangs were rampant in my community. This was already an odd stacked against me. Several of my family members got caught up in the drug war and served time in federal prison. As a young girl, this was daunting. However, my hope came from my grandmother Lillie. Like me, she was a teenage mother, and she did not let that stop her from becoming a successful serial entrepreneur. Although she was married to my grandfather, by the time she was 25, she owned land and a bar-b-q restaurant. Her grit, her grind, is what taught me to preserve when I became a teen mom.

I got the name Tisha Momma from my aunties because I was always nurturing others. At a young age, I made it my duty to ensure that others were taken care of. This was especially true when my family members' drug activities started serving them consequences. This left my sister, cousin, and I to live with our grandmother. When she died in 1993, it rocked me to my core. Not dealing with her passing of the family's matriarch, I progress through to care for myself and 21 younger cousins.

But I'm Grown

I went to a gang party in January 1995. I was 14, going on 15. A gang party is where gang members would hang out, play cards, and have fun. I wasn't in a gang, but I knew many people who were. That's where I met my boyfriend. I was sitting at the table where they were playing cards. I wasn't playing cards, talking, and observing those around me. This guy walked over and asked if he could have my seat because he wanted to play cards. I caught an attitude and hit him on the arm but gave him my seat. I was feisty, and this must have turned him on because he started asking others about me. Although I was 14, my sister told him I was 18. My mom was in jail by this time, my father wasn't there, and my grandmother had passed away. I had little guidance, so I acted like I was grown.

I was in the kitchen with my sister and noticed a girl, who was pregnant, staring at me. I told my sister she better be glad she is pregnant, or I would've smacked the mess out of this girl. I was so annoyed that I informed my sister that I was ready to go. When I said that, the guy heard me and left with his cousin, sister, and me. We went to Burger King to grab something to eat and then to my sister's apartment. The boy and I sat on the couch and talked most of the night. I was very open and told him everything that was going on in my life. We started hanging out more and getting closer. He offered something my family didn't, so I wanted more. He provided me with closeness and stability, although he didn't have everything together himself. By May, we made it official and became a couple, and I had given up my virginity. Since he believed I was 18, he didn't think I was a virgin and ultimately felt terrible for "taking" my virginity.

My real age was revealed in August when my mom was released from prison after serving 18 months. This was a confusing time in my life because my mom hadn't been there, and when she got out of jail, she wanted to resume authority over my life, but it was too late. I had gotten used to making my own decisions and doing what I wanted. I eventually moved out of the house and moved in with my sister. My mother filed a run-away police report. She came to my sister's house to bring me home and let the cat out of the bag about my age. My mom yelled at my boyfriend and said, do you know she is 15? His jaw dropped, and he started questioning me by asking why didn't I tell him the truth. I became defensive and reminded him he would drop me off at the middle school entrance and soon be 16.

Later, I found out the girl staring at me at the party was pregnant by him. They had broken up, and she was seven months. I did not know about her until I was already in too deep, and we left it at that. When I lost my virginity, I became pregnant quickly. But by the time I realized I was pregnant, I was losing the baby. I thought I was having a heavy period because I started having sex. My sister informed me that the heavy bleeding wasn't normal and took me to the emergency room. There, the nurse told me I was nine weeks pregnant but miscarried. I received the DNC procedure and went home. I cried my eyes out, and so did my boyfriend. He said that he couldn't believe that we lost our baby.

Pregnant Again

I went to visit one of my sisters in Detroit, Michigan. She kept saying, "Sis, you look like you're about to have a baby." I kept saying no, I am not. I was in such denial. But Oh, when I returned to Ohio, I received confirmation. In 1996, my mom and I visited the health department for an annual women's check-up. I had just turned 16. My mother informed the nurse that I needed to get on birth control. After collecting the urine sample, the nurse returned and said to my mom, your daughter needs prenatal care, not birth control. My mom was in complete shock. She couldn't believe that her baby was having a baby. We had to tell my boyfriend and his parents. After that, I had to tell our entire family. And it was not surprising that my family would stick by my side when they received the news that I was having a baby. My support system became even stronger.

After a while, my relationship with my boyfriend became rocky. I asked myself what I was going to do now? He already has a baby boy to take care of. I was also dealing with some unresolved issues and bad blood between him and the mother of his son. Lord, what lesson was I about to learn now! I just learned the lesson about how to take care of myself and my little sister while my mom was incarcerated and my father wasn't there. What exactly was the Lord trying to teach me about this pregnancy.

Stepping Out on Faith

I was ready for the journey of being a teenage mother. The rockiest part was getting through school with a baby, but I was not giving up. I decided I would not be a teen baby momma, raising a baby by myself.

It wasn't a part of my story I wanted to write. I saw us together forever, and I was holding on to it.

I was a cheerleader and ran on the track while pregnant. I kept asking myself if I could continue those activities, or would they come to a halt? So many things were going through my mind. All I knew was that I was bringing a tiny baby into a world where there were drugs and gangs, and a mommy and daddy argued consistently on most days.

I was only in the 10th grade. I would always ask God, why me? Why was I going through this? Some of my family members were deep in selling drugs. My boyfriend was still acting like a boy–if you know what I mean. There were so many crying nights that I held my stomach. I would always speak to my son in the womb and declare he wouldn't go through what I've been through. He would never experience being around drugs and gang violence. I remember touching my stomach and saying, "You are God's masterpiece, and I know you are going to do something great."

One day, I talked with my boyfriend's mom. I told her I knew why I had grown up so fast. I had to become the mother of my youngest sister at 13. She said that was the reason that she thought I was older than I was. I told her even though we stayed with family members, I still did all the caring for my sister; and that I kept telling myself over and over that having a baby must be a piece of cake because I had taken care of so many children growing up. We both started laughing, but I was about to find out how the baby would change my life forever. This was my "ah-ha" moment because having my baby was a distinct reality–knowing that this baby had to live with me every day and I was responsible for him- made things real. This was when I accepted my motherhood calling.

The calling wasn't easy, especially being so young. When my son was about three months old, I had to drop him off at daycare before school. That day, I was already feeling very stressed, and I received the call. "You forgot to leave your son's diapers!" the teacher scowled at me over the phone. I confessed I had left them at home, leaving out very upset, and asked if I could borrow a few from someone else. She said that I could. I remember getting off the phone and praying, Lord, please give me the strength to get through school and be the best mother I could be for this baby. I told myself that day that I would never get so stressed or overwhelmed to forget the things my son needs. From that day on, that is exactly what I did. Everything else became second, and what I needed to do for him became first.

The Real Shift

Things were going great for me. My son was doing well–he was walking and trying to talk. I was a senior in high school and was thankful to be almost finished. I said, Lord, I got through this! I was holding my own. My mom and I had gotten pretty close, so one day at her house, I said, "Mom, I cannot keep this secret." I was so scared that I was pregnant again. I just had this feeling that something was going on inside me. She said we would wait and see what happened. A few weeks later, I took a home pregnancy test, showing a very faint pink line. The faint lined had me questioning if the test was a true positive or a false negative. I yelled for my friend, and after that day, we were at the doctor's office. The doctor did a blood test. He returned to the room, lowered his glasses, and said, "positive." in a stern voice. I thought no way was I pregnant again! I graduate in a few months!

Frighten by the news; I still walked across that stage on my birthday in May. I had my beautiful baby girl in August. Now, two people depended on me, so I had to step up my game. I could not even entertain the thought of giving up anything that pertained to caring for my children. So, in 2000, my boyfriend-now husband-and I moved to Columbus, Ohio. That was where we were about to embark on a whole new journey with our little family. I was only 20 years old and was leaving my close-knit family to start fresh.

Bawse Moves, for my children!

This move is how Boss Babe came alive. I started by training underneath a lady from Venezuela who owned her childcare center. She taught me everything that I needed to know about the industry. I worked in and out of childcare centers for five years. In 2007, after going to school to get a medical degree and seeing how many girls had to drop out of college due to lack of childcare, I opened my childcare center. The steps that I took to become a boss in my industry included learning everything about the childcare world.

Steps I took in childcare.

Step One: Be teachable. I had to allow myself to be open-minded and understand that I did not know everything.

Step Two: Identify the problem. For me to find a solution, I had to know what the problem was. When I saw many women dropping out of college because of a lack of child care, I knew the solution would be to open more daycare centers.

Step Three: Don't judge others. Be willing not to judge others. So many people judged me without knowing my story. I vowed not to be that way with others.

All these steps are especially important, and that is how I became known in Columbus, Ohio as the childcare expert. Places like ECDI allowed me to teach classes on how to open a thriving childcare center. They knew me as the lady that could take a warehouse and turn it into a childcare center when others before me tried and could not succeed. I found that niche of talking to city commissioners about different buildings and learning their structures. These steps were essential because they took me to another level in the childcare industry.

The Birth of Mothers Helping Mothers Inc

What made me a full Boss and go from a Teen Mom to Trailblazer was in 2007 when I birthed Mothers Helping Mothers, Inc. This is a nonprofit helping homeless teen mothers. Taking it to another level, I could take teen mothers off the streets and give them a fresh start. That was the best part of starting the organization; while getting the attention of the United States, President Barack Obama was not so bad either. In 2016, I received the Lifetime Achievement Award from President Obama for the work that I had done in Ohio and across the country. I also earned an honorary degree in divinity. Yes, I love the Lord, and as an evangelist, I taught these girls life skills, but I also taught them the Word of God. I shared with them what it was to be grounded in the Word. Here's how I continue to inspire even more teenage mothers and young women. **1)**. Be true to yourself, accept that you entered motherhood and that you will be the best you can be at it. **2).** Make sure that you understand your children come first. Whatever you are trying to accomplish, look at your baby in their eyes and know that they should be the number one reason you want to do it and keep going. **3.)** Please keep God in the center of everything that you want to accomplish. This is especially important. **4.)** Be sure to fill your cup back up. If you›re empty, you have nothing to give.

Finally, From Mom to a Boss Babe.

1). Build and maintain a strong burning desire to do better than where I started.

2). Change your self-concept. Find out who I really was and then change the concept by seeing what I wanted for my family.

3). Start living differently. What I mean by that is I started living differently. I started thinking, talking, and acting differently and continued doing so until my life was transformed.

4). Be grateful. Many of us have a lot to be grateful for, including things we take for granted. I always make sure I think back to where I was and thank God for where I am now.

Take some time for yourself!

Taking the mother hat off for just a few hours, soaking in a hot bath, reading a great book, listening to some music, or just talking with a friend is the best advice I can give you. I want the world to know that teenage moms can be every bit as caring, loving, and perfect or imperfect as any other mother. We are parents too! We're just like other parents–only a little younger and with more to learn. On the other side of it, they can become a Boss Mom just like me. Be great!!!!

Love Boss Lady Dr. Cortesha Cowan

TEARS CAN EITHER HOLD YOU BACK OR WATER YOUR FLOWERS, YOUR CHOICE

Ty Wallace

I Missed Being a Child

My name is Tihesha Wallace, however, I prefer to be called Ty. I was born in Kansas City, Missouri, to a sixteen-year-old who already had a toddler and a drug addiction. My brother and I ended up in foster care for the early part of our lives. Upon returning to my mother's custody years later, her situation had gotten no better. She was still doing drugs, so we bounced around from house to house in drug-infested projects. My uncles lived nearby and were on drugs, so I was responsible for taking care of their children when we weren't in school and at night. When I was seven, my mother had another son, so I became a new mommy. I was the middle child but the only girl, so I suppose that is how the responsibility fell on me. My responsibilities as a mother were to fix food, have clean clothes for school, keep everyone's hair combed, and make sure we were all in bed so we could get up and do it all over again the next day.

School was challenging because the kids made fun of us, but it didn't matter because school offered breakfast, lunch, and learning. I suppose the teachers knew about my situation because my teacher called me into her class after school on my tenth birthday. She gave me ten dollars! I was so excited!! I could not remember getting anything for my birthday up until that point. I went home boasting and bragging to my cousins about the money. Unfortunately, my mother was in the other room listening. She called me into the room and told me to tell her what I was saying. I thought I was in trouble, so I was nervous. To my surprise, she was happy for me, or so it seemed. She told me to give her the money so she could buy cake and ice cream for us. I didn't want to give it to her, but I thought I'd at least get something out of the deal instead of her just taking it. It turns out; she took it for herself, and once again, I was left with nothing for my birthday. I could have cried about it, but that would have just gotten me a whipping. She never apologized or even made an excuse. My birthday is close to Christmas, and we would usually get something from the agencies that donated toys. If it weren't something she could sell, we would get to keep it, so at least I had that to look forward to.

Fast forward a few years, and somehow, I was no longer seen as a child. My addicted family members realized that the drug dealers and some of the other addicts found me attractive. I became the new means of getting drugs. "It's just a game," they would say. All they wanted me to do was wear a red nightgown for the men to look at me. It was always a red gown of some sort to draw their attention. The game evolved to more than just looking very quickly. This was the scariest game that I had ever experienced and designed by people that were supposed to love and take care of me. They told me I ruined their game by getting pregnant at fourteen. Little did they know how this little game changed my life forever.

Time to Spill the Beans

What is it like to come home and tell your loved ones you are expecting a baby, and they rejoice with gladness? I have no idea. My experience was typical of any eighth-grade girl finding out she was pregnant. I missed my cycle a few times and realized that something was probably wrong. I couldn't tell my mom, so I opted to tell my homeroom teacher, mainly because she was nice to me. She seemed to want to help me in class, so I trusted her with my situation. She bought the test to school the next day for me to take and even explained what I was supposed to do with it. She sat with me as we waited for the results and told me what the results meant and how the alternative would have looked. After a few minutes, we looked at the test. It was positive! It took me a few days to muster up the courage to tell my mother. When I did tell her, I was overwhelmed with emotion. I wanted to cry because I was so scared of her reaction, along with all the fear and uncertainty of being fourteen and pregnant. I didn't really know how to be a mother at all. I had taken care of my cousins and my brother, but what does being a good mom really require? I had no good examples around me, but I still had a deep desire to be the very best mother.

We lived in an upstairs apartment, so I tried to anticipate her reaction, especially since I'd be trapped without an easy escape route. None of the reactions that I thought of presented a positive outcome. I asked her to sit down so I could tell her something really important. She was already high by the time I had made it home, but I could not wait until she sobered up. She cussed and fussed about me taking so long and demanded me just to tell her whatever I had to say so that she could get back to doing what she was doing. I said, "Mom, I'm pregnant." She looked at me and asked how I knew, so I told her about the test I had taken. She was quiet for a while and then jumped up and shouted, "You are going to f*** everything up."

I was already crying, but her sudden movement scared me, so I really cried. I didn't know if she would hit me because she was always quick to grab any object and hit me. She has hit me with broom handles, belts, shoes, and even extension cords. This time she didn't hit me. Instead, she said, "I'm going to push you down the stairs so you can lose that baby." She gave me the meanest look and stormed out the door. I sat there for the rest of the evening, trying to figure out what I would do if she did push me down the stairs.

Two weeks went by with no spoken words, then she called me into the room. She was smiling, which never really seemed to be positive for anyone but her. She told me how excited she was to be a grandmother and how happy I was pregnant. This was a totally different response from weeks past, so now I was trying to think of what could have changed. My answer came later that day when she was talking to her brother. I heard her tell him how much more cash assistance and food stamps she would get once I had the baby. We were already barely eating and not benefiting from the cash assistance at all. I instantly saw an image of my baby walking around with dirty diapers on, no food, and crying. I didn't want that, but I had no idea how I would keep that from happening.

You may think that I could go to other family members for help, but you don't understand. My mother had four brothers, and she was the only girl. As I said, my uncles lived near us and participated in the same activities she did. However, one of my uncles lived out of town but came bearing gifts when he visited. I want to think he was unaware of what was going on, or maybe he just wanted nothing to do with the whole situation. Occasionally, my grandmother would come around and bring food from the soup kitchen where she worked. I should have been grateful; however, it was known that she didn't care for us much because we were black. Sounds strange, right? She was married to a white man and only spent time with our mixed cousins and bought things for them. When we weren't eating food from my grandmother's job, we would get our food from the church, food pantries, or worse, even the garbage.

The Game

When it was time for "the game," as my mother would call it, she would dress me up in a red nightgown, red robe, or red biking shorts and leave the room. Before she left, she would say, "It's just a game; your uncle will be here soon." I was so scared. The uncle that would show up was always someone different. Some would just have me walk around the room and look at me.

Others would actually have me sit on their laps and touch all over my body. The way the game was set up, they could do almost anything they wanted, and right before they penetrated me or took my clothes off, my uncle would come bursting into the room. He would act like he did not know what was going on and threaten to call the police. They would leave in a hurry to avoid the police. Then the game was over until the next time. If I refused, then I would get whipped with the extension cord. I had no way of defending myself against my mother or those guys, so that's where my baby's father came in. He stopped my mother and uncles from leaving me alone with those guys. Once I got pregnant, I was all on my own. I'm not sure if he felt like he failed or if he just did not care, but all I had at that point was fear and this baby. I was sad, scared, confused, and alone. Once he left, the games were free to resume without disruption, and they did. The difference now was they were free to enjoy every part of my fourteen-year-old body because I was pregnant, making it that much better.

Am I Making the Right Decision?

Our apartment was set on fire the day after the fourth of July. My family made it out of the apartment, but I was still sleeping. When I woke up, the balcony was in flames. I thought it was nighttime, and they were still shooting fireworks, so I went back to sleep. That could have ended it all. I would not have to deal with the disappointment and embarrassment however, I was really more concerned with raising a child in the same environment that I had grown up in. Wouldn't the fire be the best answer for it all? Obviously not, my mother's boyfriend ended up picking me up and carrying me out of the burning apartment.

How could I bring this child into the world and provide something great for her? I thought maybe it would be better if I just gave her up for adoption. Perhaps she would go to a wealthy family, and they could give her the life that I dreamed of giving her. I had no money, family support, or education. Who was going to help me? What were the chances of changing my situation? How could I bring this baby into a drug-infested world and only have the option of becoming a prostitute or associate with drugs to make a living? Giving up was the only logical answer.

They Believed in Me

I realized it wasn't fair to my baby for me to give up. I always thought depression was a privilege afforded only to the wealthy, so if I got in a slump, I'd just think about her future and move away from the thoughts I was thinking. She did not deserve the life I lived, and I desperately needed a big enough reason to leave my mother other than just for myself.

I had to leave my school once I started showing and go to an alternative school. When I got there, the idea of being a teenage parent became a reality. I spoke to the counselor to find out my options for success in their program and the odds of me graduating. What I found out was the first step to me bossing up. They gave me a plan, and I believed I could do it. They said if I could finish my studies early, the school would pay for me to go to the vocational college to get a trade for a decent paying job. My primary focus at that point was to have this baby and graduate early. The odds were stacked against me, but I was determined to change the outcome.

Foster Care Was a Curse that Turned into my Saving Grace

We lost everything in the apartment fire. A few agencies helped us with food and clothing vouchers, so we bought clothes for the baby and some things for ourselves. The next day, I discovered that my mother had sold everything from clothes to my iron pills. I remember thinking that she really didn't care about this baby or me if she was willing to sell the medication I needed for myself and my baby to get drugs. I don't know why I was surprised, but now I felt helpless and at my wit's end.

So I went back to the same teacher who had helped me with the pregnancy test, and I told her about my mother's drug habit and our food situation. She connected me with the school counselor, who went to talk with my brother to get his side of the story. Children's Services removed us from our home at once. I told them we had to stay together so that I could protect my brother. We went to an emergency foster home with the nicest white lady that I had ever met. She gave us our own rooms, fed us good food and even allowed us to watch TV. I figured she was only nice to us because she was getting paid.

As it got closer to my due date, I discovered how wrong I was about her intentions. My foster mom had everything planned out for what we were going to do and how things would happen once the time came to go to the hospital. Neither of us knew my daughter had her own plan; she came two weeks early. The night that I went into labor, my new foster mom had company over. I remember being in extreme pain, but I didn't want to disturb her because of my past experience interrupting my mother. I waited for her company to leave and then told her I was in a lot of pain, and at that point, my water broke. I did not know what it was, so now I was embarrassed – because not only am I crying, but now I'm standing there peeing on myself, or so I thought. She rushed me to the car and then off to the hospital we went.

She stayed there the whole time, explaining what was going to happen and what I should do next. I felt connected to her, but I was afraid to love her because I thought she might not love me back or they would take her from me. Days after we returned from the hospital, what I feared happened. The foster care system said that we had to move to another house. I did not understand why, so she explained she was just an emergency placement for foster kids. She had become a permanent foster home throughout our stay, but unfortunately, we still had to move.

My brother, daughter, and I moved to an older black couple's home. They had three other foster kids, so at least my brother had someone to play with. I was okay with the move until I realized they were exactly what I had feared the first foster home would be. They were just in it for the money. They didn't care about us or our previous situation. I asked them if they knew when we would move back to the first foster home, and they were excited to report that we would not be going back with her. I called my worker to find out what happened. I thought we did something to make her change her mind, but she said, "The foster system finds it is in the best interest of all parties if the kids and parents are of the same race." Dang!! We could not stay with her because she was white, and we were black? We ended up leaving that family a few months later and moved out of town because they feared my mother would find us. After what we had endured, I no longer believed in the foster care system. They moved us to a family that I thought was mixed. The woman looked white, and the man was black. I thought that was strange.

Well, it turned out she was black too, but I still didn't trust this new family, and they viewed me as sad and angry. One day, my foster dad said he was tired of my brother stealing, but he didn't want me disciplining him because that was their job. I told him he would not hit my brother, as that was the only form of discipline I knew. We ended up having curse communication. He said some four-letter words, and I said some four-letter words. The following day, I expected them to move us someplace else, but they didn't. We worked it out. He told me that was what family is supposed to do, and from then on, we were family. We stayed with them for two years until I got out of foster care and stayed connected until they passed.

Doing Whatever it Takes

I carried the "not good enough complex" on my shoulders like a heavyweight. I was always second to everyone else. I didn't feel worthy of happiness. I didn't think, feel, or believe anyone loved me. I was used to people only being around because they wanted something from me. So essentially, if I didn't have something to offer, then I had no one. Love was not available to me without something in return.

To transition from where I was to where I wanted to be, there were a few things I needed to do:

Choosing the MVP (most valuable person)

I had to decide that my future with my daughter was more valuable than trying to save my mother from the fate she was choosing to create for herself. My daughter was innocent and did not ask to come into this world. It was difficult to come to terms with the fact that I could not save my mother by being there.

In your situation, you may just need to choose yourself. All too often, we put others before ourselves, which includes their feelings over ours. The only person you can change is yourself; therefore, you need to focus on yourself and your emotions. It's not selfish; it's self-love!

1). Ask for Help. I had to go against my mother by telling the school what was going on at home. That probably doesn't sound bad to you, but that meant getting her in trouble. I wasn't sure if she would be sent to jail or if she would ever forgive me.

I decided I would make it work in foster care, no matter how bad it was going to be. I thought for sure it could not be worse than what I had to endure at home.So don't be afraid to ask for help. We spend so much of our time thinking about what the other person is going to say or think about us or our situation. When we let the words out, we are surprised that it wasn't as bad as we thought. Remember, we are our worst critics, so let's speak to ourselves better and find cheerleaders to help us with the process.

2). Put in the Extra Work. I buckled down and kept my head in the books. I would graduate early because the school would pay for me to take college classes if I succeeded. That meant I didn't have to figure out where the money would come from and a fast-forward shift to a better life much sooner. I didn't hang out with my classmates because I didn't want to lose focus. I couldn't fail because this was for my daughter. She needed me, and I could not let her down.

You may have to do some things you don't want to do, but those were the things that made my life change. For example, you may not feel like working an extra shift, enrolling in school, or hitting the gym, but if that is what you have to do to get ahead, how much do you want to change?

3). Get Out of Your Comfort Zone. We had to leave the foster home because my brother was giving my foster parents' problems. They were supposed to move us together, but they separated us. I had to get out of foster care so that I could get custody of my brother. I worked on getting emancipated. I was old enough to work, go to school, and take care of my daughter, so it was time to get my own place.

Changing our lives can be scary, but don't allow that to stop you. When you have to do something that makes you uncomfortable, then just follow your feet. Sometimes, if we can just keep moving a little bit at a time, we will be surprised how close our goal really is.

4). Get a Plan. I did it!! I graduated and got myself out of foster care. Next, I had to figure out what I wanted to be when I grew up? I knew nothing about professions except that being a doctor or a lawyer would make you rich or at least more prosperous than I had ever seen. I already knew that I couldn't be a doctor, but I still wanted to be in healthcare. I could become a CNA at the vocational school and work my way up to being a nurse. Yes, that sounded like a good life for us.

Make a plan for our life. If you fail to plan then you plan to fail. Even if you don't reach your planned goal, you will always be much closer than you were when you had no plan at all. If you don't know where to start, start at your wildest dream and work your way to something reasonable. Either way, you can do it!

5.) Allow Room for Change. Next, I found love, had my second daughter, and got married at nineteen, divorced by twenty-one, and now a single parent to two children. I always wanted to find someone to love me and live a good life. I thought I found that until he started using the same drugs that I grew up around. I couldn't allow my girls to grow up in the same environment I had worked so hard to escape.

Unfortunately, there will be monsters from your past that will reappear. It is up to you to make room for them in your life or shield yourself from those past mistakes. They will appear shiny and new, but remember where you are going and evaluate if the past will help or hinder your growth.

6). Know Your Worth. I got so fed up with people using me and not contributing to my growth that I moved to Texas. A friend said I could get into nursing school easier in Texas after being rejected two years in a row. I just wanted to have a good career to provide a better life for my girls. I had previously worked for the railroad and was laid off, so I needed to fix my financial future quickly.

Too often, we wear the labels that others put on us. Most of the time, they are projecting their feelings about themselves onto you. Don't fall for it. People try to remind you of the old you if you have changed, but don't fall for that either. You are amazing and can only get more amazing as you learn and grow.

7). Make it Work. I applied for any program that my prerequisites qualified me for because I didn't waste any more time. I needed a career! I received a letter congratulating me for getting accepted into the

Radiologic Technology Program. I remember not knowing what it was at all, but I was convinced that I would complete it successfully.

Life will give us the unexpected, and it is up to you to say yes to the opportunities. Just because you don't know what something is, give it a try anyway. It could be the best thing for your life.

8): Never Stop Learning. After completing the program, I continued my education. I took additional courses to advance in radiology; then I thought it would be cool to be the first person in my family to have a bachelor's degree. I also sought mentors to help me along my journey.

We look at life backward at times. We go to school to learn how to do certain things, but we think we don't need anyone to show us in life. That could not be further from the truth. If you are trying to get somewhere, having someone who has already made it there will be invaluable.

Triumph through Adversity

I did it!! There were struggles along the way, but we made it. I never made it to becoming a Nurse, but I have one of the best professions where I am part of a leadership team. I am proud to say that I have changed the future for my children and their children. I worked hard to ensure that they did not live in the ghetto, go hungry, or go without. I worked two jobs once I did complete my training, and I worked even harder to advance. During that time, I missed out on events, get-togethers, and we even celebrated holidays late, but it was not a problem. We turned it into a tradition and got good deals too! I achieved my most significant accomplishment! My children did not have children while they were children. My oldest daughter had "My Honey", who is my grandson, when she was twenty-three years old. She graduated from high school then realized college was not for her. She is now an aspiring entrepreneur. My younger daughter graduated high school, became a successful athlete – resulting in a full-ride scholarship to a great college, and graduated with a biochemistry degree. I'm so excited to see how life turns out for them. Knowing that the sacrifices I made gave them a hand-up so that they never needed a hand-out is comforting to me. I am always looking for new ways to challenge myself and to take my life to new heights. I believe in mentorship, which has helped me a great deal in my professional life. I am currently a relationship coach, and I am also working to get my time freedom back by being a successful investor. I believe that you have to keep learning and growing and I work at that intentionally.

Challenge yourself to learn something new every day and apply it to your life. You will be surprised where you will be a year or so from there.

You Can Do Great Things

I did not know how I was going to make it out of my situation. I was scared. I was from the "ghetto", so I was already prejudged. I had no one to show me anything aside from my everyday reality. How did I even think that I could be better? I guess I figured that anything had to be better than the life I was living.

I have been asked several times how I made it out without becoming a drug addict, alcoholic, criminal, or worse. The only logical answer that I can give is that it was for the glory of God. He planted this burning desire to be, do, and have more – even after accomplishing a goal. I pray, believe in myself and the goal that I am going after, and keep going even when things get scary. I like to read, so my imagination allows me to see my dream coming true before it happens, which fuels me to keep going.

I recall hearing people at the level that I'm working towards say things like, "If I can do it, so can you, or it's simple to do but not easy." I never believe them until I get frustrated than do the work and achieve the outcome. Our minds are so funny. We will listen to other people to confirm what we are fearing is true, but the truth is, you can do whatever you set your mind to do. Just believe and keep moving forward. If someone, like a coach or mentor, tells you to do something specific towards your goal, then just believe in their belief and do it.

If you have found yourself in a situation similar to mine, don't listen to those negative thoughts. You have something great to offer the world and this new life you are bringing into the world. If you are dedicated to being the best and to continued learning, so you grow into even more greatness, then anything is possible. When I'm scared or unclear of the direction I'm going, I'll stop and remind myself that everything works out for my good.

I invite you to practice the same thing. Everything works out for your good. Say it with me. Everything works out for my good! Now that your mind is in the right place, sit down with a piece of paper. Write down where you currently are in your life and where your wildest imagination could take you. If it scares you, then it is a great start. If not, then keep writing until you have no idea how the goal can become a reality. Now break it into smaller chunks that you can see yourself accomplishing alone or with help.

Asking for help was the most challenging part for me. I felt like no one cared about my situation, but the fact is they just didn't know to care. People may not know that you need help either until you give them the opportunity to help you. That could happen in many different forms. They could help give you a new perspective on your situation, give you encouraging words, connect you with the resources to get help faster, or a myriad of other ways. You don't have to tackle this on your own.

I have to end this portion of the book by saying that I am so happy and grateful to have been a teen mom. It taught me how strong I really am, and my background gave me a blueprint of the life that I did not want for my girls. They were the fuel to our life's greatness. Your baby can be the same, something amazing that will push you forward. It's okay to cry; just know that your tears will water the flowers that you will enjoy for the rest of your life. Love your baby, forgive yourself, and go after the life of your dreams.

Boss Lady Ty Wallace

FROM FALLING APART TO FALLING INTO PLACE

Dr.Tanikka C. Price, EdD, JD

Promise to God

I am Dr. Tanikka C. Price, Esq.- that's right, I am a Doctor and a Lawyer! I have a law degree and a Doctorate in Education. I grew up in Linden, on the Northside of Columbus, Ohio. My parents met and married in college when they were teens, and they have different stories about why they married and whether they planned me in love or was a mistake. My mom said she wanted me, but my dad made it known that he did not want children (me). Growing up, there were times when I felt like I was wanted and times when I felt I was a burden. I understood how a parent could feel like that when I found out that I was pregnant as a sophomore in college.

I was not ready to have a child, and I did not want to be tied to my boyfriend, Logan, for life. Laying on that cold table during my two-day procedure, I looked up at the light, and with tears running down my cheeks, I promised God I would never make the decision to abort again. When Logan suggested abortion, I was ambivalent. He didn't see any other way. However, it wasn't until the nurses at my college infirmary made the appointment at the abortion clinic for me that I realized I had chosen to terminate that pregnancy. I didn't even ask my child's father to be there for me during the procedure- I knew I couldn't depend on him. But I didn't expect my best friend to abandon me the morning of the abortion. I had to make the hour-plus drive by myself from Poughkeepsie to White Plains. I was numb, and I made my mind up. I kept telling myself that I was not ready to have a baby; I was still trying to convince myself. During the 12 weeks that I was pregnant, nightmares haunted me about neglecting and abandoning my child. I thought I would inflict the same trauma on my child that had been done to me. I had one recurring nightmare: I would leave my child in their crib and party at the club. Another recurring dream was I tucked my baby into my pocket to hide her because I was ashamed. These dreams and the fact that I never wanted to have children because of the trauma I experienced in childhood led me to face these issues surrounding motherhood in counseling shortly after my abortion.

When I was four years old, a cousin molested me. That same weekend, my mother told me she and my father were getting divorced. My mother's anger and disappointment was projected on me, and I bore the brunt of a lot of her frustrated fists and sad slaps on the face. I walked through my childhood holding my breath and waiting for the other shoe to drop. That unhealed and unresolved trauma led to a domestic violence relationship in high school, in which I was raped, and then my relationship with Logan, who grew up in a New England suburb far from the hood in Ohio. We had an on-again, off-again relationship that was unhealthy from the beginning: drugs, hip-hop, and clubbing mixed with some cheating and lies made for a toxic mix. I was so distraught about feeling pressured into the abortion that we broke up but, because we were comfortable with each other, we continued a sexual relationship. Although I was using birth control, I became a teenage mom to my daughter, Kalia, six months later. This is the story of how my pregnancies changed my life, giving me scars that I am still healing from today while ultimately giving me the fuel I needed to prove naysayers wrong and become a Boss Mom.

A Teenage Love

Not wanting to rely on my 46-year-old memory to write about my experience from 25 years ago, I looked back through my old journals to gather the story of my pregnancy. I've been writing and keeping journals since I was in 5th grade. I learned that year that writing was a way to help get me through the dark places, a place where I could be safe and free. I searched through six or seven journals until I finally located the one I kept during my pregnancy. It was one of the smallest notebooks among my journals, and I separated it into four pieces. I think it was so small because I hid my pregnancy for a while initially, so writing about it was my little secret. My first journal entry was 6/15/1995, and it read:

> "Guess what? I'm pregnant, and I'm keeping it. I can't believe it... I want my baby. Of course, Logan wanted me to have an abortion, but I'd rather have it. I think it's going to be a boy. I don't mind either way, as long as it's healthy and happy. I just hope Logan can be here for us. But if not, I'll be okay. I have many options."

This was so true. I don't know how I was so sure and so confident that my future would be promising, considering I had no job and wasn't finished with school. I was in college the first time that I found out that I was pregnant. The voice of the nurse that gave me the pregnancy test results at the campus clinic, the voice of my boyfriend, and the voices of my friends were all that I heard in my head. It was just like that lyric in Lauryn Hill's song *Zion. "Look at your career, they said, Lauryn baby, use your head."* Mine was more like- look at this scholarship! You cannot go back to the hood!

They outnumbered me! I didn't believe in abortion because two of my best friends had been born to teen moms, so I had seen that sacrificing for your child was beneficial in so many ways. But the pressure to keep my scholarship and the questions that Logan was constantly asking me about my faith: "How do you know God is real?" "Why would God let babies die?" "Where was God when you were being molested?" My faith was unraveling, and so was my relationship with God. I drove over an hour from my campus to the abortion clinic when the nurses had scheduled my appointment. I will never forget the sound of being in that room. It overwhelmed me with many emotions: guilt, relief, sadness, happiness to have my body back, regret, and I stuffed the feelings down for many years. After the elation wore off, I found myself depressed and overwhelmed with my decision. I am still processing the unforgiveness I've been carrying towards Logan and myself. We knew we weren't for each other: he wasn't faithful or honest, and I was trying to make him into something he wasn't. We argued and disagreed on so much, but we stayed connected because we had sexual chemistry. Honestly, we were probably trauma bonding- only I was willing to talk about mine, and he never did. Having an abortion, combined with being in a toxic relationship and using recreational drugs, took a toll, and I found myself in a suicidal mind state. The grief and guilt were so heavy that I remember crying and banging on my bedroom floor, asking the earth to swallow me up one night. I didn't have the strength to go through with a suicide attempt or end my relationship, so not much changed in my life. In the age of AIDS and STIs, I'd rather remain with the same person than start all over again with someone new. I was afraid of getting hurt and scared of being vulnerable. I stayed miserable and continued my "relationship" with Logan, using drugs to dull the pain.

Six months later, on summer break at my mom's house, due to my daily addiction to "...cannabis sativa, otherwise known as the chronic, I didn't even notice my period was late. As a weedhead, I had stopped paying attention to details. My period was synced with my college roommate's, so I relied on her to tell me when it was time for me to buy pads. I was having what I thought was premenstrual syndrome (PMS) symptoms: sore breasts and cravings for Frosted Flakes and Rice Krispy Treats. I was also exhausted, but that came with the territory of regular cannabis use. I called my college roommate to ask her when I should expect my period. She was the responsible one who tracked the periods for both of us since we were in sync. It shocked me when she informed me she had not had one but two periods since we left campus! It floored me when the clinic physician did an ultrasound after my positive home pregnancy test and told me I was 12 weeks pregnant! He told me I only had a few days to decide whether to keep the baby or terminate the pregnancy before abortion was no longer a legal option. This time was different. I wasn't on campus with the campus nurses scheduling my abortion and telling me I would be right back to normal. Logan wasn't there pressuring me with what people would say or think, and I vividly recalled the promise I made to God on that cold table staring up at that bright light. I knew precisely that I had to do what I had to do.

Our last sexual encounter had been "goodbye sex" because Logan got kicked out of our college because of poor grades, and we assumed we would finally be done with each other. After that last time, I bent on letting go of the toxic relationship and focus on myself. Little did I know that there was going to be more than me to focus on forEVA! I couldn't believe it when I found out I was pregnant and how far along I was! When I told Logan, he asked me to get another abortion and said that his last girlfriend in high school had several abortions and was fine-I wasn't her- I told him I committed to keep my baby and raise her myself. Logan and I disagreed, and I thought he would convince me to terminate again. While my mom and grandmother were supporting and loving, his parents were disappointed and judgmental. I was in my third trimester of the pregnancy before he said the words *"...when we have the baby..."* rather than *"...if we have the baby."* It was good for me to find out I was pregnant during the summer. Not being around Logan physically gave me the space to center on what I really wanted for my life.

My pregnancy was idyllic. I felt connected to my baby from the beginning. I captured in my notebook that I had one of the scariest things happen to me during the summer. I was working for Camp Mariah Carey as a music instructor for inner-city kids in New York. Mariah Carey visited the camp one day, and my choir sang "Hero." Everybody clapped when the campers sang the high note from the diaphragm, just like I had taught them. Shortly afterward, I had experienced sharp period pains and had to be rushed to the hospital. When I first laid on the hospital bed, tears ran down my cheek because the pain felt just like the pain of the abortion. I thought maybe I was being punished for what I did. I didn't want a miscarriage, but I believed everything happens for a reason. I spoke to Logan, to let him know we might lose the baby. I could tell that he was also very nervous. He wanted nothing bad to happen, and he wanted his child to be healthy." In the end, I was just dehydrated. After some fluids and admonitions to drink more fluids, I went back to camp. Although I didn't have a miscarriage, I'm still dealing with those feelings of guilt from my abortion- because I had done something that was against my value set.

I had finally made it to the end of my pregnancy. When I went into labor at my mom's house in Long Island, NY, Logan was at his parent's house. I had just driven over two hours each way to see him. Although our relationship was rocky, we had great sexual chemistry when things were going well. I think it was that physical "connection" at nine months pregnant that brought on my labor. I drove the 2 hours back home with my water broken. It wasn't a gush, like in the movies and TV shows, but trickles. I would never have known it was my water breaking if my Granny hadn't noticed me waddling back and forth to the bathroom every 5 minutes once I got back home. She said, "Nikka, what are you doing?" I said, "I keep peeing on myself!" Granny said, "Nikka, your water probably broke. We better head to the hospital." As always, Granny was right. My mom was there waiting when my daughter was born. When the Doctor lifted her up, I saw Logan's whole face. I thought, "Noooooo, I'm going to have to look at Logan's face for the rest of my life!" I can laugh about it now. I see him in her face daily, but my first thought when she was born made me want to cry.

Even with that, she was the most beautiful baby that I had ever seen. It took her a while to open her eyes. Shortly after looking at her, it was like someone whispered her name in my ear: "Kalia." I was feeling so many emotions- Logan wasn't there, as usual, and that set the tone for how things would be for years to come. I didn't believe that I could make something so perfect. My mom held her, but I did not want to hold her, partially because she looked like Logan–and nothing like me–and partially because of the unexpected pain of the C-section that I had to have because the baby was breech. The doctors kept thinking she would turn her head down, but she never did. She sat up like Buddah, and I had to be cut open. That pain was like nothing I'd ever experienced before. Hours after I'd had her, when I was still refusing to hold her, one nurse waited until my mom left and then rolled Kalia's bassinet next to my bed and left. It was late in the evening, and when she started crying, I realized there was no one else to get her but me. I scooped her up in my arms for the first time; she was so tiny and so perfect. She looked at me with those big eyes, and immediately her crying stopped. When she stopped crying, my tears started falling down my cheeks. I realized she was all mine, and love rose in my heart that was bigger than any emotion I'd ever felt. These tears falling were not coming from fear but tears of love and joy. Kalia was a sign of hope, restoration, and God's immense grace.

I'm Going Down

The hardest part of being pregnant at a young age was dealing with other people's pressure on me. First, not having the support of my child's father had such an enormous impact on me. That was the hardest thing that I went through. I already broke up before I got pregnant. He wasn't there to support me when I had the abortion, and when we tried to make it work during my pregnancy, he cheated on me. The day I went into labor, I couldn't find him (this was before cell phones), and I found out later that he was with another girl-after I'd just driven two hours to be with him. Later, I'd find out that one girl he was with was my so-called friend during my pregnancy. Talk about trust issues! I built walls around my heart because of this relationship that took years of therapy and prayer to remove. I still feel uneasy with my single friends around my husband, thanks to this situation!

Logan was ashamed of my pregnancy, ashamed of me, and ashamed that he got kicked out of school. I refused to take on that shame. Whenever I felt as though I wanted to give up on school, I thought about the strength of my ancestors and all they had to endure as enslaved Africans in America. My minor in college was Africana Studies, so I studied the obstacles, perseverance, and fortitude that our ancestors had to develop to survive slavery. It also helped me think about my 80-year-old self: which decision could I live with when I was old and reflect on the essential things in life? I am so glad that I could look beyond my teenage self to my 80-year-old self to lean on my value set to make my decision. I thought at the time that I was choosing my child over my education and future opportunities, and I knew that at 80, I'd rather have my child than my degree. Thank God I never had to choose and could have my children and complete my education.

What You Won't Do for Love

My friends, male and female, were so supportive during my pregnancy. They stopped clubbing, and we would all sit in my dorm eating fast food and playing cards. That meant so much to me. My friends helped me take care of Kalia, and they matured and got serious about life right along with me. My turning point for staying in school came during a conversation with one of my friends, Terrence Oliver. I told him that having my baby meant I was choosing to forego my full scholarship to college. I would have to drop out, leaving my dreams of being a rapper, writer, and educator, and go back home to live with my Mom and Grandma. Terrence, the wise soul he was at such a young age, said, “What would you tell your daughter to do if she were in this situation?” I said, “I’d tell her to stay in school, you can do anything you put your mind to, and God will make a way.” He said, “She will never do it unless you do it first.” That was the moment that changed it all for me, my Oprah Winfrey “a-ha moment.” I have been thriving ever since because I decided that day to do it first so she would know she could do it too. Where there is a will, there is a way. Let no one make you feel like your life is over because you have a child. This can be the beginning of a different, more meaningful life.

Girl on Fire

It wasn’t easy to transition from falling apart to things falling in place, but the choice to be better for me and my children outweighs the option to remain the same. So, here are the steps that I took to BOSS UP!

1. ***Know who you are.*** Your decisions should always align with who you are and your personal value set, regardless of how you were raised or what church, mosque, or temple you attend. Certain decisions just aren’t you. As soon as you recognize that, regulate and get back to you. Before Logan, I’d always been a church girl and never used alcohol or drugs. His lack of faith affected me and had me questioning God’s entire existence. Once I doubted God, it was an easy jump from questioning God to drug use, and from that to an abortion, and then to addiction to numb the pain. All that happened in the three years we dated. Moving away from God, I started making choices that didn’t reflect who I was or what I believed, and eventually, I no longer recognized myself.

2. ***Find some solitude before big decisions or changes.*** I found my way back to God through Kalia's eyes. In her eyes, I saw everything good about the World. I saw both God and my own reflection. It helped me to realize that I was good and that I was worthy of forgiveness. The way back to God began at Camp Mariah when I was alone and pregnant. Many of God's soldiers have significant alone time. Think of Jesus–when He was alone in the Garden of Gethsemane. Think Joseph–when he was in jail. Think Moses–when he was wandering in the desert. That alone time allows you to get quiet and figure out who you are and what's best for you. My quiet time came during the summer of 1995 when I spent the summer at The Fresh Air Fund's Camp Mariah. I woke up in a cabin by a lake, in the middle of nowhere; this was before cell phones, so no one was calling. I could reach out and make calls when I had time–and the emotional capacity–to deal with other people's opinions. In that quiet place in nature, I could feel God's presence, and I knew I was not and never would be alone.

3. ***Pick and choose your inner circle.*** Be careful of wolves in sheep's clothing. I had a best friend who I thought had my back, but I kept hearing that she was talking about me, and she kept leaving me hanging when I needed her most. She was supposed to drive me to my abortion appointment but canceled the morning of the appointment, saying that "it wasn't her responsibility." Wow! My 15-year-old cousin went with me and drove me home–she is my ride or die. Your friends and your sexual partners will have a massive influence on your life, so pick with caution. Sharing bodies with someone will have you sharing their view of life. Friends will expose you to either the best or worst that life offers. Make sure you choose friends

that lift you up. And remember, you don't have to let everyone who knocks on life's door inside.

4. ***Pray and never stop praying.*** Regardless of how you grew up– there is a spiritual force that is bigger than us that we must depend on when we are weak. Nothing about life is easy, and it's impossible to make it without God's strength. It is only with the unconditional love and acceptance that comes with knowing God that you will be able to see yourself through life's traumas and tragedies. Talk to God just like you talk to anyone else, and He will make himself known to you. Seek, and ye shall find, knock, and the door shall be opened unto you.

Everything Is Everything

"A Baby is God's sign that life should go on." I had a sign with that proclamation written above my bed while I was pregnant with Kalia. It inspired me in the hard times. I've often said that I made my kids with God. I prayed over my stomach when I was pregnant and dedicated all of my children to God, regardless of the circumstances in which they were conceived. He stepped in and made a way–time after time because I trusted Him completely. When Logan's parents asked me how I thought it equipped me to take care of a baby with no money or education, I said, "God will help me." They were not impressed or amused, but God did it. People always donated clothes and baby items to me from the time I was pregnant onward. I would be in the grocery store, and someone would say, "I have baby clothes in my car; do you want them?" Miracles happened to my babies and me repeatedly. You cannot do it alone. You can create a support group–a family that will love and accept you where you are. Pregnancy is a time of tremendous growth and change. Capitalize on this time to create the life you imagine. Remember, if you believe you can–you're right. If you believe you can't–you are also right. Your kids will do it if they see you do it first.

Started from The Bottom, Now I'm Here

I stayed in school and graduated in 5 years from Vassar College. I thought I was the first woman to have a baby and continue her education. Although I was the first to have a baby on campus openly, I found out that another strong, black female had a baby at Vassar and graduated before me. I stood on her shoulders. I attended law school–attempting 1L (year one) twice and ultimately graduating. I took the bar exam three times, and finally, four years after I graduated from law school, I became a lawyer! Right now, I'm living the life that I imagined. I have six kids, and I've been married for thirteen years. This is not the only story in my life, but this is the one that changed everything. Having Kalia Monique was the best decision I ever made after giving my life to Christ at age seven. The Lord has sustained us, and although I made some poor decisions and found myself in unpleasant situations after this, I never wavered in my faith ever again. I stood firm on my belief and asked God to forgive me each time I fell. God has been the rock and the light that has guided me since I decided to keep my baby. That baby that I stepped out on faith and had is now a 25-year-old nurse who is married to her high school sweetheart and is a mother to two beautiful children. That's right; I'm a Mimi to Mila and Malachi! Like the great poet Langston Hughes said, "Life ain't been no crystal stair," but I have never again taken my life or my children's lives for granted.

Love Boss Lady, Dr. Tanikka Price, Esq.

GROWING UP IN THE FAST LANE

Ciera' R. Hale

How it all started

When people ask me, "where are you from?" I laugh on the inside because I consider myself a suburban girl, with a pinch of hood … lol! To sum it up, I'm an east sider born and raised in Columbus, Ohio. My grandparents lived in Berwick. However, my mom lived on Elaine Road, which has never been a good area. My upbringing was kind of different, not traditional at all. My grandmother worked a 9-5, and my grandfather cooked, cleaned, did my hair, and made sure I was ready for school. Few families in my community were like this. I was 9 or 10 when my grandmother passed away from cancer. After that, everything changed.

My First Heartbreaks

My mom left for the army. My dad was addicted to drugs and alcohol, which led him to prison. My grandfather remarried, and I was left to go live with my aunt. My aunt provided for my cousin and me. She made sure we were taken care of. She held down two jobs for as long as I can remember, which kind of left us to fend for ourselves. I was a young girl with no accurate guidance or direction, which led me to grow up so fast and down the path of teenage pregnancy. They all were my first heartbreaks.

Innocent or Not?

I was your average 12-year-old playing with my best friend, riding bikes, or playing with our 20 Barbie dolls on the front porch. Still, I was a little more mature than everyone that I was around. It seemed to have come naturally. My best friend says that I was the momma of the group – not even knowing or realizing that small statement would turn into reality at thirteen. I was hanging out with my cousin and some of her friends. One guy, who was her cousin on her dad's side, was interested in me. I guess I would say we liked each other. He eventually became my typical middle school boyfriend. We would only see each other on the weekends because we didn't go to the same school or live in the same neighborhood. One day, I went over to his house with a group of friends. We were lying on the couch in the living room watching television. He was behind me, and I was lying on my back. He rolled me on my side and lifted my blue jean shorts and underwear to the over. That quick, we engaged in sexual intercourse, and no one around us noticed. I pretended as if nothing happened. The next day, I remember being over at my friends' house playing Barbie Dolls, as if nothing ever happened.

Then on another day, as I lay across my bed, I got extremely sick. I took two giant steps towards the restroom and then vomited everywhere. I immediately knew that I was pregnant. That was the first and only time of the nine months of carrying my baby that I ever had morning sickness. Missing my menstrual cycles, even that young, was a clue to let something inside of me know I might be pregnant. I was scared, and I knew that something was wrong, but I didn't say a word to anybody.

After that, I moved to El Paso, Texas, with my mom on the military base where she was stationed. Days, weeks, and months went by; despite that, no one knew I was pregnant. I continued with my daily life by going to school every day. In my mind, others could see the growing bump in my belly, but the baby's growth scared me. I would often punch myself in the stomach with extreme force. I wanted it to all stop! I wanted this moving thing out of my stomach. In my young mind, I wondered, how could I have gotten pregnant fully dressed?! In the movies, it didn't go down like that. I hid this pregnancy as if my life depended on it. All that I could remember was my mom telling me like it was yesterday, "Girl if you ever get pregnant young, I'm gonna kill you." I took her extremely seriously. I was about seven months pregnant, and I started having these weird spells where I would pass out in class. My mom took me to the doctor, and they asked for a urine sample. I had to figure out what to do so the secret would remain safe. I added water to the urinalysis cup. I only added a little, but enough to make the sample come back negative or unreadable. They told my mom that I was just dehydrated, and my mom still never thought that I was pregnant. She would ask me if I was still having my cycles, and my answer would always be yes, knowing that I had not had a cycle in months and that I had a suitcase full of pads in my room.

I knew dressing like a tomboy every single day and wearing a jacket in 100-degree weather to hide my pregnancy would soon end. The Memorial Day weekend of 1999 was the day. We were on our way to a cookout, and we were all sitting in the car getting ready to leave. This day it was so boiling hot that I had to take the purple jacket off. When I took the jacket off, my mom could see my poked-out stomach, and the secret I kept to myself had ended. She looked over and asked in the softest tone, "Ciera, when was the last time you had your period?" I looked up at her and said that I don't know, with tears in my eyes because I was so scared. I thought I was going to die at that second. She said, "Ciera, what do you mean you don't know?" She then said, "Oh my God, Ciera, are you pregnant?!?!? Let me see your stomach. Unbutton those pants; they are wayyyy too tight. I took off my jacket and unbuttoned my pants in shame to uncover that I was indeed pregnant. Oh no, Ciera, why didn't you say anything? Why didn't you tell me?" I cried my eyes out. "Mommy, I'm so sorry. I'm sorry. I'm so scared…." Not just because I was scared at that moment, but I didn't know what was next. At that very moment, I felt something I hadn't felt in a long time. For as long as I could remember, my mom seemed more of a friend and not a nurturing mommy every child needed and wanted. At that moment, she gave me everything I needed and wanted, the thing I longed for more than anything. It was just so simple. I needed my mommy.

Who's The Baby Daddy?

I gave birth to my son on June 11th of 1999. I was in labor for 13.5 hours, birthing a 7lb 11oz baby boy at 13. I hid my entire pregnancy with no prenatal care or doctor's visits at all. Surely, I thought I would not make it. I was so scared. I did not know what I would do if that day with my mom wouldn't have happened. My son's father told my mom that the baby wasn't his, that I was lying, and that he didn't know what I was talking about. The only emotion that I could feel was an indescribable feeling of fear.

That summer, I moved back to Ohio from Texas with a baby that seemed to come out of nowhere. My family and friends couldn't believe that this child was mine. I lost a lot of friends. No one wanted their child to be around me because I had a baby. I felt so ashamed and alone. Fall came around so quickly that year, which meant school was starting, and I had to get a babysitter. Thankfully, my Great Aunt agreed to watch my baby so that I could go to school. I would wake up early and walk my son to my aunt's house every morning before I walked to school. I was drained and didn't know how I was going to do this every single day. It was already hard enough to take care of this baby. Once again, my mom was left for the military, now overseas, at a base in Seoul, Korea. I just wanted to throw in the towel. I couldn't see any light at the end of the tunnel.

Can I Right My Wrongs?

I was 17 years old, almost 18, and my son was ready to go to kindergarten in the fall. I was all over the place. I was supposed to have graduated that summer with all my classmates, but I didn't. That sent me to a dark place, into a depression. I started running the streets and moved out of my aunt's house without a plan. I went to live with my cousin. I ended up sending my son to live with my mom, who had returned home from Korea, to go to school. I had no means of income. I woke up every day with no plan, no vision, nothing at all. Yet, this 5-year-old who called me mommy had no clue what I was going through. I always made our struggle fun for him so the truth wouldn't affect him. I then realized that I had to do something different. I wanted my son to come home. I never wanted him to be away from me. So I changed the trajectory of my life. I got my GED, got a job, and had an apartment. I was still very unsure about what I wanted out of life. I kept pushing forward, and my new boyfriend kept pushing me to do better. He stood by me and my son's side. He became the father that my son didn't have. He stepped right in.

The Cycle Breaks HERE!!!

I was pregnant at 12 and gave birth to my firstborn at 13. But listen, that didn't stop me, even though I surely thought many days that it would. I earned my GED and an associate degree in Medical Assisting and became a serial entrepreneur. I started working at a daycare center, and I loved it! I had this underlying love and passion for people and for kids that flourished. When I saw the center was owned and operated by people of my similar background, right then, while working hard for that company, a light bulb instantly came on in my mind, and I knew what I wanted to do. I didn't know how or when–but I knew one day, someway, somehow; I was going to own my childcare center.

The Process

Although I was extremely young when I was pregnant, life taught me some significant lessons. Instead of being bitter and upset, I created a path to help you transition from teen/young mother to boss lady!

1. Adjust. I had to understand that this was my life and no one else's, so the things I could no longer do, I just could no longer do– such as hang out with my friends whenever I wanted to. Therefore, I had to adjust to my new life and not be resentful towards myself.

2. ***Realize it's not about me.*** I had to understand that not only was I still growing up, but that I now had a whole person who looked up to me for everything that he needed, so failure was NOT an option. I had to focus on the bigger picture.

3. ***Don't take it personally.*** I had to understand that people's judgments were their judgments and that just because everybody has an opinion; doesn't mean their views have any relevance to me or my life.

4. ***Mistakes don't mean I'm a mistake.*** I had to accept that I could make mistakes and still correct them and move forward with whatever I wanted to do. I'm not perfect; I'm human.

These steps played a vital part in becoming who I am today, and I didn't even realize who I was for a long time. I had to step back and analyze all that I've overcome. I had a baby young, and I made mistakes. In fact, the mistakes didn't end with me having a baby at 13; I've made plenty more. But God is the orchestrator of my story, and I will own it and be the boss that He created me to be in the end.

At 27 I opened my first childcare center - Loving Hands Learning Center, LLC in 2013. Then I moved on to launching a childcare consulting service, 1st Kids Inc., where I help others make their dreams of becoming childcare owners come true. I now have a second center, which is home-based – New Beginnings Learning Center of Columbus. Might I add I love the kids, and it's one of my biggest passions. I love watching them come into my center as infants and phase out of my program as school-agers. That's such a joy! Yet, your boss lady didn't stop there. I've now moved on to a hidden talent I've had and have been sitting on, letting it marinate for the right time. I've become a licensed nail technician – Cee Me Shine Nails by Cee. I also started my non-profit organization, S.H.E is ME (Standing Here Encouraged), where I empower young women to push through even when life says something different. Look at me! I'm still pushing and never stopping. It's UP from here! Might I add that baby boy is now 22-years old, has graduated high school, and has started his own detailing company. He has made me a proud grandmother to my Gran Gran Man! I now have two more sons, one daughter, and a bonus son, who are all such a blessing to my life.

To my ladies and maybe gents, God never makes mistakes. Everything that you're going through, or have gone through, He knows the story first-hand from the beginning pages to the ending chapter. Call on our creator when you feel like giving up and feel like there is nothing left to do. When you have those sleepless nights, and your mind is racing because you have mouths to feed and don't know what's next, pull deep down on the inside and grab hold of what we call the silver lining and push like never before.

There is a purpose in every one of us. There is a BOSS deep down on the inside of you, Sis. Your BOSS is waiting on her grand appearance to the world, as well as to the most important person – HERSELF!!!!! Don't let what the statistics say about us teenage mothers win or negatively affect you. Beat those odds and boss up! I did, so can you!!!!!!

With love, Boss Lady Ciera'

FROM CHASING LOVE TO LOVING MYSELF

Ruby Cole

Confused love

I grew up in a house that was not a stranger to abuse. From the outside looking in, our house probably looked peaceful. We always had nice cars, took vacations, and had the latest technologies. There were always cookouts, parties, and celebrations. However, on the inside, there was everything but peace. There was arguing and cheating. Hiding in closets became a typical Friday night thing when my parents would fight. Seeing this at such an early age had me confused. I did not know nor understand love. I saw my mom in so much pain, and I often wondered why we never left. Sometimes we would leave, but that only lasted a couple of days, then we were right back. I promised myself that my love would be different. I spent most of my childhood looking for love.

Breaking Free

Once I was old enough to go on my own, I was never home. I was always hanging with friends and stayed out as late as possible. My best friends and I started going into a town about fifteen minutes from ours. The boys interested in us hung out at a local barbershop. The barbershop was the place to be. People constantly were in and out, joking and talking to each other. After going there a few times, this guy started showing interest in me. We exchanged numbers and started talking on the phone every night for hours. Eager to be with this boy, I started going to that town every chance I got so that I could hang out with him.

One day, I let him convince me to have sex. I don't know if we ever became a couple because the truth was; I had a boyfriend at school. I was in a dilemma because I liked my boyfriend and this guy. They were complete opposites. My boyfriend was a good guy, played football, and got good grades. My mom loved him. The other guy-let's just say he was not a stranger to school suspensions. I kept seeing both juggling my time between the two. I was only having sex with one of them. When I got pregnant, I was 17 years old and a senior in high school. I did not know that I was pregnant, but my mom sure did. I remember we were riding down the street, and she hit the side of the door and said, "you need a baby like you need a hole in your head." I was so lost and confused by her comment. I never noticed the change in my body or realized that I did not have periods. She later took me to the doctor, where we found out that I was nine weeks pregnant. At the time, I did not know how to process this information or what I was going to do.

My mom was incredibly supportive. She even gave me the option to have an abortion. I thought that the baby's father and I would be a family, so I kept the baby. I based my decision to keep the baby on what I thought love was. I thought by me having his baby; he would be madly in love with me. Little did I know how far from reality that would be. The guy I was pregnant by was also a teenager. Me being pregnant did not stop him from doing his dirt. He kept living his life while I struggled day-to-day carrying his child. He did not neglect me, nor did he show me any loving signs. This was hard for me because I was still on my search for what love looked like. He kept seeing other girls, making me feel even more inadequate and less loved.

Being pregnant made it hard to keep up with school. I was tired all the time and often fell asleep during classes. I started skipping school and only showing up for the essential classes. It was so hard to focus on school, but by the Grace of God, I graduated. Six months pregnant, but I made it. I could not go to college like most of my friends and that hurt. I wanted the entire college experience. Instead of orientation and moving in dorms, I spent my time chasing after love.

Finally, my son was born and reality set in. I was 18 years old, fresh out of high school, with a newborn and a job working at Burger King. I felt so embarrassed but did not think I could have lived with myself if I had gotten an abortion. I lived in a small town where everyone knew everyone. Going to the local grocery store was like a family or class reunion. So, I would go in and out of stores very quickly to avoid talking to people. Of course, my close friends and family were very supportive. Most of my friends were continuing their education to better their lives. And here I was, lost in a system of food stamps and broken promises.

Numb Love

At some point, the fighting between my parents stopped at home, but it left a negative impression on my life. My mom was a shell of her former self. Miserable in a life of lies and left betrayal. Yet, she still loved the source. We were just living a numb life, going through the motions. My mom was excellent with my son, and she was my backbone. Her love for us allowed me the freedom to still be a teenager. I could work, go to a community college, and still hang out with my friends. It appeared I was going to be alright. However, that changed when my mom became sick. The hospitalizations became frequent. I leaned on my son's family once this happened. Then tragedy hit; my mom passed away at 42. I was 21 and devastated. I felt like I was losing my entire world, my backbone, my strength. No one could understand my pain. I was heartbroken. The only person who genuinely loved me was now gone. The search for love continued, but this time it was worse because now I was trying to fill the gap of my mother's love.

Looking for love in all the wrong places

Thank God for my ex's family. They would keep my son for days and weeks at a time. I knew he was safe, and no harm would come to him because they loved him so deeply. He was their joy, and this made me happy. No one knew I was slowly spiraling out of control. I was looking for love in all the wrong places. I did an excellent job of hiding my actions and masking my pain. Sex became my addiction, my new love. Whether they were married, single, or divorced, I did not care. That type of love would only last for one night. Then, I would wake up empty again, searching for my next fix. Why was love so hard to find? I latched on to any bad habit that I could find, smoking, drinking, and sex. All these things only offered temporary relief and had me instantly searching for more. I had dropped out of community college so I could hang out more. I was living life in the fast lane, reckless and free. Meanwhile, I still had a whole toddler to care for. Some days, I did not know how I made it. I would literally go from the club to work. No one knew the truth, so I thought.

Somebody's Always Watching

My destructive behavior forced me into my turning point. Someone told my uncle about how my life was going. He, in turn, told my aunt, who lived in a different state. My aunt convinced me to move to Ohio. It was not until I moved to another state and spent time with my son that I realized my life was miserable. I had no dreams of a future. Here I was with a 3-year-old and did not even know what tomorrow held. It had been so long since I cared for this child that I did not know how to function. I had to start over. At times, I felt like giving up. I remember I would think to myself, 'I'm going just to quit. I'm going to move to the projects and live off food stamps and welfare." Then I would look at my son and decide that he needed and deserved better. Because despite my emotional lack of love, I was well taken care of. Growing up, I wanted nothing. How did I get here? I remember thinking that I was a failure. Thinking, how was I going to recover from this? I was my mom's only child, so I did not even have any brothers or sisters to lean on. My relationship with my stepdad was non-existent. He had moved on to another family. I lost my mom and my stepdad at the same time. My life had changed completely; nothing was the same. I had lived my life for 21 years, and now I was lost. I felt like someone had just dropped me off in the middle of nowhere and said, "Good Luck."

The Turnaround

I think once I realized things were not as impossible as they seemed, I began to thrive. I found a job and got an apartment. Things started to look better, but I was still heartbroken. Just going through the motions of life, doing what I thought I was supposed to do. Meanwhile, I still had that empty feeling and no proper direction in my life. I met a guy, and we started dating. He moved in with me, and I finally started feeling some sort of normalcy in my life. He loved and helped me care for my son. Although I appeared happy, something was still missing. There was still that gap that I could not fill, no matter what I did. I ended up pregnant with my second son at 26 yrs. old and again thought that I had found love. We broke up after my second son turned 2. After we broke up, I vowed never to be in a relationship again. I was done looking for love. Now it was time to work on me. Self-love was what I needed. I needed to heal and transform myself. I started going to church and reading the bible more. I also learned the importance of allowing God to lead me. Make a plan.

Do research.

Never give up.

There will always be bumps in your path. What you do with those moments sets you apart from others. You can give up or push forward. Focus on your goals.

My Recipe to Success.

There were moments in my life where I acted out of pain instead of my genuine authenticity. I allowed the pain to outshine my creativity. Therefore, I must show you how to shine your light!

1.) **Recognize your talent.** I recognized how talented I was. Once I stopped chasing love and started paying attention to myself, I realized I could do anything I set my mind to do. I was great at crafting and decorating. It also helped to calm me; it was very therapeutic. When I was crafting, it felt so natural and unforced. I had loved myself more.

2.) **Tap into your skills.** Once I started tapping into those skills by studying and expanding them. I started buying books and crafting all the time. I would spend hours perfecting whatever I was working on. People loved what I was producing, so I started investing more time and money into my activities. Loving yourself has got to be the most important thing you can do.

3.) **Determine your directions.** I had to decide which direction I was going to take. Was I going to start a job involving my skills or start my business? People were asking me to create things and paying me good money to do so. Once I received requests to decorate weddings and birthday parties, I knew I had arrived. I had found something I was good at and enjoyed doing.

4.) **Write your goals.** I wrote out my goals and plan. After realizing that I could make a side hustle out of event planning, I planned how things would operate. I had a good clientele, and they were constantly referring to my services. I was working at an event every weekend. With the encouragement of friends and family, I started a successful event planning business. I was so good at what I did; I created the outfit, catered for the food, and decorated for the event.

5.) **Put God first.** The most important thing I did was put God first. I have done much praying over my life, but I was not listening to the Lord. God will talk to you and direct your path if you let Him.

The Life God Had for Me

Currently, I am a credit center specialist. I had not even been working for the bank a year when a bank manager approached me. She said that I did excellent work, and she could tell that I was a good person. She said that there was a position coming up that she wanted me to have, and the rest is history. I am now a serial entrepreneur; I have had many side hustles, such as an event planning business, made tutus, and I teach others how to coupon to put more money in their pockets and not in stores. Currently, I am the founder and owner of Agitation Custom Crafts, specializing in tees, drink-ware, and trays. I pride myself on bringing people's ideas to life. I am also a life coach specializing in empowering women to find their success, no matter their age. I am starting the S.H.E initiative. S.H.E meaning Smart Honorable Excellence. Leading middle-aged women in finding their genuine passions and teaching them it is never too late. I moved to Alabama at 42 and started my life completely over. I have thrived and been successful in everything that I have done so far. If you walk with confidence, confidence will follow.

Words of Advice

Do not be afraid to invest in yourself. If possible, remove yourself from toxic atmospheres. Look at your life and figure out what you can change. Then move forward and do not stop until what you could change is behind you. Do not be afraid to change. You may lose some friends along the way, and in that case, they were not your real friends, anyway.

Surround yourself with like-minded people. When you are around people that have ambition and dreams, you feed off of each other. Remember, iron sharpens iron. If you take this advice and apply it to your life, I promise you will win.

Boss Lady Ruby

FROM HIDDEN SECRETS TO MANIFESTED DREAMS

Darnisha Davis

God's Plans

Life is full of ups, downs, and surprises. I didn't know that my life would turn out how it did, and let's just say, I am grateful. I didn't grow up with a silver spoon. I lived in a single-parented home and did not have a relationship with my biological father until a little later in life. My mother did the best that she could with what she had. She kept us in a strict apostolic church. I am the second of four girls, now women. They considered me the quiet one. I didn't say or complain much. I learned to deal with problems internally, including sexual abuse. Sometimes, I would wait too long to discuss things that troubled me. I ended up a teen mom at age 14. Although I thought my life was over, God had other plans. I became a licensed hair braider, and now, I am the owner of Shay Lovely Braids, LLC. I also have a vending machine business with my oldest child, whom I became a teen mom. God changed my life for the better, and I am so happy that I didn't give up.

The Church Boy

Life caught me by surprise. Although I was the quiet one, I tried to keep up with my oldest sister. We were inseparable. Wherever she went, I wanted to go. Well, on this particular day, my sister went to her boyfriend's house and guess what I did, I followed her. Her boyfriend had an adopted younger brother that was there as well. We were introduced to them at church. We were all a part of a singing quartet group. I had known him a couple of years prior. I really didn't like him. At 14 years old, I didn't know what my type was, but he wasn't it. My sister and her boyfriend left us sitting in the living room together, and we started talking. He asked if I wanted to go to his room. At this time, I was a virgin, and I was afraid out of my mind. I didn't feel pressure because I was emulating my older sister. She was with her boyfriend, so I was cool with being with him. He was 15 years old, and we honestly didn't know what we were doing. I had heard stories of the "cherry getting popped," but I didn't experience that, nor did I know if he "finished."

After this sexual encounter, I didn't talk to the church boy much. We were in the age of no cell phones or social media, so I would only see him when my sister went over his house or at church. As close as my sister and I were, I didn't tell her he and I had sex. I continued to live life as usual until I realized I missed a menstrual cycle. This led me to tell my oldest sister because I was seeking advice and scared. She told me we had to get a test to see if I was pregnant. We were so afraid to tell my mom; we devised a plan. We told our mother that we were going to check on Mother W from the church. We knew that if we told her we were going to Mother W's house, she wouldn't question our whereabouts. We went to CVS Pharmacy to get a pregnancy test and stopped at Mother W's house to check on her, but primarily to use her bathroom to take the pregnancy test. I anxiously sat on the toilet and peed on the stick. I was staring at the test waiting, hoping it was negative and, surprisingly, the two lines appear... "Dang, I'm pregnant!" I called my sister in the bathroom, and we both panicked and were not sure what we were going to do. We weren't sure how, or if, we were going to tell our mom, and I didn't want to disappoint her.

I was terrified. I didn't know what to do or think, but I knew I didn't want an abortion. Although I was only 14, abortion was not an option with my church background, and if my mom found out, she wasn't taking me to an abortion clinic. I called my child's father and told him the news. He inquired about an abortion, but was supportive of my decision not to kill our child. I was a little relieved. I was still in denial; I went on with my life as if I wasn't pregnant. To distract my mom, I used my older sister's maxi pads. My mom was like an FBI agent when it came to me and my sisters' menstrual cycles. She knew when we started and how many pads we used each month. If anything was off, she questioned us.

One time, we took a pad and put ketchup on it to throw her off. I would also wear baggy clothes to hide the secret.

Secret Revealed!

The craziest thing happened that got me caught up. It was my mother's wedding rehearsal, and I was a bridesmaid. It was a dress rehearsal, and I had to try on my bridesmaid dress, six months pregnant. I had to walk down the aisle, not realizing that my belly button was poking out. Nobody said anything until the end of rehearsal. Once I went to the bathroom to take off my dress, my mother had a pregnancy test there waiting for me and demanded that I take it in front of her. To this day, I don't know where she got this pregnancy test from.

My heart sunk. I knew I could no longer hide my baby. All the hiding convinced me I wasn't pregnant and, at that moment, my pregnancy was real. I even went to an amusement park when I was six months pregnant. Of course, the test came back positive, and it was just like I received the news all over again. I saw my mother cry, and that was worse than anything in the world. I was so disappointed in myself because of the decision that I had made. I felt embarrassed and disappointed in myself. The next day she took me to our family doctor, and the first thing the doctor said was that it had better be a tumor and that she was disappointed. I had lost so much weight due to not having any prenatal care. I went from weighing 135 lbs to 109 lbs. I thought if I starved myself and didn't eat, the baby would vanish in my mind. I was so severely malnourished that I ended up breaking my tailbone because of a lack of nutrition. I literally bent over and heard something pop.

It was difficult starting high school and being pregnant. I was teased and talked about for being pregnant in the 9th grade. I remember walking into my classroom, and a particular girl would say, "Wow! You're in the ninth grade and pregnant? You're a hoe", while the entire class laughed in the background. There were days I was just so embarrassed and ashamed that I wanted to walk out of the school, but I just kept it to myself and kept pushing. I had to encourage myself. It was hard, but I knew I had to for my soon-to-be baby girl.

Ready or Not

I had a 36-week checkup and ended up receiving a scary surprise. The doctors diagnosed me with preeclampsia. I had toxemia and high blood pressure that could fatally harm the baby and me. They told me to go home, eat something, and then prepare for my labor to be induced. I remember being home but not being able to find anything to eat. My mother boiled some chicken and noodles that she had gathered. I went back to be induced, and I was in labor for 24 hours. All I could remember was feeling like I was starving. I was crying so badly, and my lips were so dry and chapped that they were cracking. However, the hospital staff let me get a couple of licks of a popsicle. I ended up having a C-section because of my baby girl's heart rate continuing to drop. Finally, my baby girl made it into this world. I wasn't sure how I would take care of her, but I was glad that she made it safe.

After the delivery, the craziness did not end. The exposure to the sexual abuse that I had experienced finally came to light. It had torn my family apart, and I was just trying to manage being a teen mom. The sexual abuse started when I was 12 years old by a close family member. There was chatter about whether my baby girl was the product of the sexual abuse, but fortunately, that was not the case. The pressure was so bad that we picked up the pieces of our life and moved to Columbus, Ohio. I was on government assistance while attending high school, and I had the help of my mother. She never left my side.

Although my life had some wrong turns, I turned my tragedy into triumph with the help of my support system and God. I graduated with my diploma and became a STNA (state-tested nursing assistant). I moved out of the house at age 18 and became a wife. Although the marriage ended in divorce after five years, I was blessed with a son. Even though that was not my plan, I let nothing stop me. I continued to work and take care of my family at all costs. Let me share with you what steps I took to manifested dreams.

1). **Overcome shame and guilt.** Shame and guilt plague me like a scene in a horror movie. Every time I would try to run from it, it was right behind me. The guilt I felt was from the weight of becoming a teen mom. The shame I felt was this painful feeling that I dishonored my mom and myself. The reality was there was nothing I could do to change the fact that I was pregnant at 14, no matter how many times I tried to deny it mentally.

2). **Heal from early childhood trauma:** There were several things I had to heal from, but I specifically focused on the sexual abuse and church hurt. Our upbringing, environment, and experiences shape us. If we're not careful to heal from them, we will carry them on to our children.

3). **Build your confidence**: Many situations in my upbringing caused me to think little of myself and not feel worthy. Therefore, I had to work extra hard to build the courage to know I was enough. I did this often by affirming myself. I practiced speaking positively over my life every day. After a while, it shifted my mindset.

4). **Forgive others.** My childhood trauma wasn't my fault, and I didn't deserve to be taken advantage of early. However, it was my responsibility to release what happened to me. When I forgave, it freed me. I didn't feel like I was bondage anymore.

5). **Consistently show up.** I had to constantly show up for myself and my children. No matter how I was feeling, what I had to do, or what was or wasn't going on in my world. If I gave up, I wasn't just quitting on myself. I was quitting on someone who didn't ask to be born. I made it my priority to finish school, start a business, and be the best mother I could be. Showing up doesn't mean that you'll always get it right. Showing up means you never stop trying.

Nothing Can Stop Me!

I realized that life happens, and you must live with the choices you decide to make, good or bad. I pushed through. I can't tell all of my story in this chapter, but know that there is more. My tenacity pushed me to become a licensed hair braider and the CEO of Shay Lovely Braids, LLC. My and my now three children want for nothing, only by God's power, grace, and mercy. The baby girl that pushed me into my destiny is now grown, has graduated high school, and has started her own business. I know that if I had given up, I would've been writing a sad story, but this is a story of victory.

Let no one tell you what you will become. Know for yourself that you are more than a conqueror and that you can achieve anything if you put God first. Stay focused, and keep going. I am thankful for all that I have been through because it has made me the BOSS I am today. Keep your head up and if you get nothing else out of what I have shared, understand that you have the strength in you to complete your assignment and destiny.

Love Boss Lady, Shay

MAKING YOUNG MOTHERHOOD A BLESSING NOT A BURDEN

Tasha M. Yarborough, CPA, MAFM, MBA

Two-Time Teen Mom

I was a two-time teen mother. I had my first daughter when I was 17 and my second daughter when I was 18. I was on birth control, but the birth control failed. I hate to admit it, but it most likely failed from user error. I would miss days or fail to let my biological father know when I needed a new prescription. Mostly, I had the mindset that getting pregnant would not happen to me. My boyfriend, at the time, later disclosed that he was getting me pregnant on purpose so that he wouldn't lose me.

I was able to hide my first pregnancy for many months. It started with me being naive, and then moved into straight up denial. I was a student athlete, captain of my drill team, and member of the all-star city drill & cheer team. As an all-star, we practiced for months and then traveled to a competition in Florida. I was pregnant during all those months of practicing the award-winning choreography. Even crazier, my biological father got the bright idea to drive my baby sister and I down to the Freaknik festival in Atlanta, Georgia and left us there with his sister. So, I was pregnant, holding on to this secret, and was in the middle of this crazy way-too-old for me festival. That was a very scary experience. My sister suspected that I was pregnant and touched my stomach when she thought I was asleep. That may have been what shot me into taking action. I decided to talk to my mom when I finally got home from Atlanta.

The responses that I received from the news that I was pregnant varied. I was living with my biological father but told my mom first. She told me that she already knew and was just waiting for me to say something. After telling her, I told my biological father. When he found out, he said that I wasn't the first and I would not be the last. I went back to live with my mom after that. The most reactive response was my high school. My school held an assembly on teenage pregnancy upon learning of mine. Me getting pregnant was a big deal. I was the model student – I was smart and had a 4.0 GPA, I was taking advanced placement courses, I was in the national honor society, I was an Ohio State University Young Scholar with a full ride scholarship to college at OSU, I was the drill team captain, I never had any disciplinary issues, I had many friends and treated my classmates with respect – I was darn near perfect school wise. The shock and wakeup call was understandable, it had become clear that teenage pregnancy could happen to anyone.

Shame vs. Support

I had gone months without pre-natal care. After getting connected with a private obstetrician, I had a racist experience. During one of my ultrasounds, the racist obstetrician remarked that my daughter had Fat Albert soup cooler lips. I was so upset and couldn't believe that in 1995, he had actually fixed his mouth to say that to me!

Teenage pregnancy can come with a lot of embarrassment and shame. I felt mortified in the doctor's office that day because of his ignorance, but I felt embarrassment at school for another reason. On the first day of my senior year, I was ready for the school year. I have always loved school, when I was little, I used to inquire of my mom when summer vacation would end and when school would resume. That day my hair was freshly styled from the hairdresser, and I had on a light pink floral sundress with spaghetti straps. I looked beautiful, I also looked all of nine months pregnant. I looked like I had swallowed a basketball and it felt like everyone was staring at me for all the wrong reasons. I was so embarrassed and full of shame that I decided not to return to school the next day. I went on maternity leave and the school district sent me a homeschool teacher for home instruction. Up until then, I didn't even know that maternity leave in high school was a thing.

Another source of embarrassment and stress can come from having a lack of finances for the things that you need, and want, from the lack of having a job. I still recall the embarrassment that I experienced from showing up to a friend's baby shower with a very inexpensive gift (it was a single onesie that had only cost a couple of dollars, in a gift bag without tissue paper). I didn't have much money, so I got what I could afford. There I was, showing up pregnant myself with an insulting gift. The father's reaction told the story of just how pitiful the gift was, and everyone trying to hush him, made it worse.

Just like embarrassment, shame, and stress are real, so is true family and true friendship! During my first pregnancy, that homeschool teacher was very impressed with me. She recognized that I had potential and she encouraged me to continue with my education and dreams. My community of high school friends were also very supportive. They all came up to the hospital to support me during both my false labor (during a Friday night football game), and my real labor (that Sunday night). They calmly told me that they had to go to school in the morning, so I needed to go ahead and have the baby, lol! She decided to follow suit and entered into the world that night.

My oldest sister and I were pregnant at the same time. I had my first born in September and there were many days until that January when she took care of her allowing me to sleep-in until late morning. It gave her a chance to care for a child, as she would later die from childbirth complications and never be able to care for her own child in that manner. Losing my sister was sad and difficult, but at the time, I didn't mourn much because I knew that I would see her again one day. It wasn't until years later when the heaviness of the tragedy fully weighed on me.

After I had my first, and later my second child, my mother and stepfather (who is my self-proclaimed father without the "step" prefix) would drive me across the city to my classes at The Ohio State University and watch my children while I was taking classes. They would also watch my nephew since my sister had died as a result of childbirth at the age of 19 and they were raising him. My oldest daughter, nephew, and second born daughter, all grew up close like siblings instead of cousins. They were stairsteps in age, being born in 1995, 1996, and 1997. My close-knit, spiritually grounded family was a huge part of my success.

Sacrifices vs. Blessings

Going to college was a given for me. I had a full-ride scholarship since the 6th grade, so even after having two children, my mind never wavered from that expectation. I decided to stay in Columbus to keep my support system. Had I not become a young mother, I would have likely gone to Spellman College, a prominent HBCU (historically black college/university) in Atlanta, Georgia. That was one of the major concessions that I had to make. There were many things that I had to forego and miss out on because I was pregnant including missing my last Young Scholars' summer institute, not being able to take driver's education in school so I had to delay getting my driver's license until age 18, and so on. However, just like the sacrifices are real, the benefits can be real too. People don't much talk about the positives of being a teen parent, but there are some. Besides the obvious that it matures you at an accelerated pace and may keep you out of the streets from getting into mindless trouble, there are other positives. Having children young made me eligible as an "Independent" student per the free application for federal student aid (FAFSA) for college funding, I was able to use my own income versus my parent's and I was able to rely on my own tax paperwork. This was something that I considered to be a blessing.

Another positive was that my body was more cooperative than when I was older. My first two pregnancies and labors were relatively easy and uneventful. That was not the case with my son (@ age 25) and third daughter (@ age 37). My two latter children almost did not make it, and they almost took me out with them. Even historically, having kids young is not as taboo as it is in our current society. Back then it was not a social epidemic or something to overcome. In an agricultural society, reproduction has more pros than cons. However, in our modern society, it can make life for the children, parents, and family more challenging. So, in general, people try to discourage teenagers from getting themselves into that situation. Nonetheless, if you are a teenage parent, we, the authors of this book, support you and are sharing our stories with you as a means of encouragement, hope, support, and examples that it can be done successfully.

Family Drama

I haven't discussed my oldest daughters' father much, other than that he was intentionally getting me pregnant. There was a lot of drama between my family and his family. I was young, impressionable, and I felt caught in the middle. Our families have a great relationship now, but it wasn't always that way. I think tension between families can be common. Emotions can run raw when people are volleying to make sure that their children are taken care of, and not taken advantage of. As far as he went, God Himself broke up our relationship. I refused to be a baby factory. God had bigger plans for me, and I was pushing the limits by staying with that person. After we broke-up, he had a specific idea of family, the nuclear family. So, when I decided, well when God decided that I no longer would be in a relationship with him, he decided that he no longer wanted to be a father. If he couldn't have me and us all living together in one home, then he refused to continue with his role as father to our little girls.

I raised my children as a single mother. I made the decisions, I made the schedules work, I scrapped together the coins needed for mortgage, food, school lunch, school clothes & shoes, activities, medical, everything. I didn't know just how self-reliant that I had been all those years, until now. I got married in 2020, and my eyes finally opened to the option of having another person to help pick-up the children from daycare or figure out the budget.

Unfortunately, there is a dangerous reality that I have to speak to. When people have children young and decide to part ways, they are highly likely to move on with other people. Those new people may not respect your children's boundaries as well as they should. This is an important warning to heed. Be careful and get to know people extremely well before trusting them to be around your children. Make sure that you listen to your children, be aware of changes in their behaviors, and check them for physical signs of abuse.

I grew up impoverished, in a home seeing physical, sexual, drug, and alcohol abuse. It is interesting that I previously didn't connect these negative life-shaping ills with having an impact on my life, but looking back, they may have shaped some of my choices. I had an interesting and very diverse upbringing. Yes, there was the home/food insecurity and the abuse, but there was also an extremely strong spiritual foundation and plenty of love and guidance. I maintained a strong sense of self and looked for acceptance from within, versus trying to find acceptance from external sources. In my mind, I did not skip a beat after getting pregnant. Having children was not an automatic stop to any of my pre-set plans to go to college and to have a successful future. Failure did not enter into my mind and did not shape my mindset. I am appreciative of my support system that enabled me to continue without skipping a beat. I would have been successful regardless, but it was much easier with my family's support.

I had so much confidence in my ability to make it no matter what, that I was actually happy when the county government kicked me off of welfare. It signaled to me that it was up to me to meet my needs. I relied on God, the Father, and he always made a way for me and for my children. While in College, I secured an internship at a local retirement system making $20/hour. I was able to buy a new car with basically zero miles on it, and become a homeowner at age 21 when I bought a newly built condominium. I was doing it! While I was doing it and things were going great, I tripped up and got involved with someone that I had no business being with. After a couple of years, I got pregnant again with my third child. With my two children in my teen years, I had no thoughts, nor did I have any pressure from anyone else, to have an abortion. However, when I got pregnant as a 25-year-old young adult while being in a rocky relationship with a 34-year-old, he emphatically stressed that I have an abortion. I told him that I would have my child with or without him, and I did. While I regret being in that relationship and all of the hell that ensued, I do not regret having my son.

Let's breakdown six must-dos for success:

1. ***Have faith in God*** – I firmly believe that there is a most high God that grants me the breath of life, orders my steps, and deserves my praise. When challenges are getting the best of me, it is comforting to know that a higher power exists and can work it all out for me.
2. ***Seek a support network*** – Imagine moving from one house to another by yourself. Now, imagine how move day would be exponentially easier if a few friends showed up to help you out. That is exactly how life with and without a support network feels. Get you a crew that you can rely on for childcare, transportation, and finances, if you can.
3. ***Dig deep inside yourself*** for the confidence needed to achieve any goal or conquer any challenge – You can literally manifest

something from nothing if you believe that you can. Take on any challenge, small or large, with the mindset that you will win if you put in the work … and then put the work in.

4. ***Take time to think and reflect often*** – People are usually able to give great advice to others, but not to themselves. Some of that is because we don't stop to step through our own issues using an objective lens. On a regular basis, get in a quiet place and tune everything out so that you can just be in the moment and think. Go with whatever decision brings you peace.
5. ***Visualize your success*** – Creating a vision board really works, it keeps your goals literally before your eyes, which keeps them on your mind. Also, being clear about what you want communicates to the universe what it is assigned to do to help make your goals reality.
6. ***Remind yourself of your wins*** – We can be so hard on ourselves because we are focused-in on achieving goals that we haven't yet reached. This feeling of discontentment serves as a motivation to keep going. However, it is important that we also gain the fuel and confidence that we need to go on from revisiting our previous wins. Remind yourself of how awesome you are, it does wonders for your esteem to believe that if you have done something great once, that you can do it again!

Love Begets Love … and Success

I had my fourth and final child at the age of 37, I mention this child because my pregnancy with her exemplified what can be done when you have grit. During my pregnancy, I was working a full-time job, establishing two start-up businesses, serving on the board of a local organization, taking courses to get my second master's degree, and studying for the Uniform Certified Public Accountant (CPA) Examination, which is a NO JOKE – hard as hell, four-part exam! On top of all that, I was successfully managing gestational diabetes, my mental health, and successfully mothering three other children.

There are only 365/6 days in a year, so everyone shares a birthday with someone else. During two of my pregnancies, I took notice of famous people that were birthing their children into this world at the same time as me, and I took note that my children were just as special to God as those kids were. During one of my hospital stays, a local billionaire and his wife were down the hall of the labor and delivery unit. I was moved to a deluxe room because the halls were sectioned off and abnormally secured because of the high-profile birth. When I was in labor with my son, I kept hearing about Labron and Savannah having their first son, my son desires to be a professional athlete and has the heart and drive of a true champion. Whatever his future, he is equally loved by God. Know that you and your child(ren) are also highly loved by God and have a purpose to fulfil on this earth.

At the time of this writing, my oldest child is in college and owns her own candle business, Jace & Mace Candle Co. She didn't become a mother until her twenties. My second oldest child is in her senior year of nursing school. She is a volunteer for social causes such as the prevention of, and healing from, sex trafficking, and she is a future entrepreneur. My son is a high school junior and has a high aptitude for all things electronic. My third daughter is extremely bright, and caring. At age four, she became a 4-year-old CEO when she started her own bookmark company, Brooklyn's Bookmarks, LLC. She is now five years old and heading to kindergarten.

I couldn't be any more proud of my children or myself, we all are living examples of God's grace and glory. We all continue to boss up in our own special ways, and I have been blessed to marry a boss husband. He courageously stepped out on faith in the middle of the 2020 pandemic when he left a W-2 job and started two businesses, YBG Environmental Services and YBG Transportation Services, under the parent company Yarbrough Business Group, LLC. I have seen him work hard to earn his personal training license and dream big with future business ideas. The year 2020 was bittersweet for many people. While we lost some family and friends, we were also able to realize many major milestones, including building a house from the foundation up and getting married. My story has gone from being a two-time teenage mother, to being a boss! During my journey, I have always been a force to be reckoned with, and I will continue to be because of God, my family, and my inner strength. You too have an inner strength. Just the mere fact that you are taking time to invest in yourself by reading this book shows that. You are seeking answers and I hope that I have helped to illuminate some keys to success for you.

Love, Boss Lady Tasha M. Yarborough

KNOWING I AM ROYALTY

Sharita Davis, RN

Lonely Girl

I grew up in a middle-class household of a single mother. I was the only child of my parents, who divorced after years of spousal abuse. My mother worked 2 or 3 jobs to ensure that we could live in "nice" neighborhoods. We lived in 2-bedroom apartments for many years. I was often left alone as my mother spent most of her days traveling to her jobs to make ends meet. I often felt lonely and hated being alone. Left with household chores, I grew to loathe washing dishes, folding clothes, and making dinner for myself. My mother monitored me via scheduled phone calls and pop-up visits on her lunch breaks. Many of her jobs were second shift or overnight. She would often sleep during the day to prepare for work. She was present, but I still felt alone when she was home.

By the age of 14, I was awaiting the birth of my first child. Finally, I had something to love and keep me company. Finally, I had someone who would cure the loneliness. I had a baby! This baby was going to solve all my problems. This would ensure her father would be present forever, and love me forever.

How was I to become a doctor now? My family was expecting me to graduate and become Dr. Sharita Davis, MD. Fortunately, my story does not end with the birth of my baby. I graduated from Temple University with a bachelor's degree in biochemistry. I later returned and obtained another bachelor of science in nursing. I became a best-selling author and the proud owner of Raiment by Rita, LLC, an online medical apparel store.

Girl Meets Boy

I met my boyfriend through a friend who was her boyfriend's friend. We met by telephone first. We would talk for hours almost all night long. I would talk to him until the sun came up. I would be so tired the next day, but I was so excited that he liked me. One night he asked, "Where is your mom?" I had been content talking to him on the telephone all night. I never thought about asking him to come to my house. I was excited about spending the night with my love. In my mind, we would talk and laugh all night long, just as we did on the telephone. I invited him over to my house while my mom was at work. To my surprise, the expectation was more than talking.

That night, I had my first sexual encounter. It was filled with innocence, fear, excitement, and void of pleasure. I was still excited to say I had been with him because I developed deep feelings for him. I felt he would never leave my side. We had several more encounters that ended in a missed menstrual and possible pregnancy.

I went with my girlfriend to Planned Parenthood to confirm what I already knew. I was pregnant! The fear that filled my mind upon learning this truth was overwhelming. I didn't cry. I didn't yell. In fact, I had no emotion until I went home. Once I reached home and was alone, I cried. Fear-filled questions bombarded my mind. What had I done? I was afraid of my mother's reaction. What was she going to do or say? I had let her down. She was going to "kill me." My grandmother was going to be so disappointed. I had shamed my family. All these thoughts and realities filled my head, and I still had not told my boyfriend. What was he going to say? Would he truly stick around? Would he marry me, although we were not old enough to get married? Where would we live? I was working as a server after school. He was working at Burger King. Did we have enough money to support a baby? What would his family think?

I got enough courage to call him. The phone rang, and my heart was beating so fast. I had images of movies in my mind. Would it be as theatrical as the movies? I hesitantly told him I had something to tell him. Afraid how he would respond, but he guessed it right away! He was happy. He was excited that I was having his baby. We agreed to hide it until we could tell "the mommies."

I continued to work, but my uniform started getting tight in the middle section. I didn't go to the doctor's office to get a check-up; I thought I was ok. Then it happened. We went shopping for school clothes. My mother always purchased new clothes each school year. She would have me try them on when she got home to ensure each outfit fit appropriately. I tried on a piece of clothing that must have revealed my growing belly and caught mom's attention. She said, "go take it off AND come back out here. You have something to tell me." I gasped. That was the longest walk of my life. My immediate thought was... she knows!

I walked back into the living room. She told me to have a seat. She asked, "are you pregnant?" I denied it. She asked again. I denied it again. I had been putting sanitary pads into the trash all these months to disguise my pregnancy. I thought I cured her suspicion. I stopped denying my pregnancy when my mom said, "if you are pregnant, you need to see a doctor." My concern for my baby outweighed my fear of my mother. I needed to see a doctor to make sure my baby was ok.

Upon my admission, my mother asked, "who is the father?" I was humiliated at that question. How could she not know? I told her his name. Her response was to call his mother. So we did. His mother's response was neutral. His mother said she needed to speak to her son. My boyfriend called me and said, "So the mommies know?"

My mother told him he needed to "get up here as soon as possible." He came the next day. He was so brave. He faced her. My mother told him to "go on about his business." He told my mother, "I'm not leaving my baby." He was very supportive. He was present for every doctor's appointment. I didn't have a baby shower, but he purchased every item that he promised. The physical part of pregnancy was exciting. I remember the first time I felt my baby move. I was so excited. I was going to be a mom! I called her dad, and he sounded excited too. I felt so complete the first time he placed his hand on my stomach and felt our baby move. I finally had a family! I finally felt loved.

One day, I was home, and the phone rang. I answered, and the voice on the other end asked to speak to me. It was a female voice. She told me she was also pregnant by my boyfriend. I replied, "Why are you calling me? You need to call him." It devastated me. My world had collapsed with that one call. I called him, and confronted him about the call. He denied even knowing the girl. My suspicion of his involvement lasted man years, but there was never any confirmation

He was with me during the delivery. He watched the monitor with each contraction. He laughed as I pulled my hair out. I yelled. I screamed. I cried, but finally, at 1:26 pm, I delivered the most beautiful baby girl. She was born with her eyes open and smacking her lips. She had a bald head. No hair! He rushed to see if her eyes were green like my mother's eyes. He told my mom that she was going to have my skin color. My mom said he yelled out the window to each of his friends as she drove him home. He screamed, "It's a girl! I had a girl!"

I missed my prom because of the pregnancy, but we agreed to go to his prom. When his prom arrived, he said he was taking another girl. I was crushed. Summer came, and he told me, "I don't have time to play house with you." He would take the train to my house and cross the street with pampers for our baby, but the visits became fewer during the summer. Our baby would wait in the window for daddy to come across the street. My baby and I spent many days looking out that window, crying because he did not come across the street. Many future relationships failed because of the disappointment that I felt during that time. I never trusted men the same. I felt all men would fail me eventually because of this disappointment.

My mom and my grandparents were the keys to my success. I could continue working and attending school. My mom cared for my daughter while I attended school during the day. My grandparents would care for my baby while I worked. We had the dream team. I graduated from high school with honors. I got accepted to Temple University with a full scholarship. I graduated in the top 10% of my high school graduating class.

With all my scholarly achievements, I still faced so many emotional challenges. I was scarred. My heart had been broken. My dreams of marrying my daughter's father had failed. He had a new girlfriend, and she was having his baby. I had a new boyfriend too, but my heart had been damaged. I saw nothing beautiful in my reflection. I had disappointed my family by getting pregnant out of wedlock. I had disappointed my daughter by not keeping her father in her life. I often wondered who would want me with a baby?

I spent most of my twenties trying to find myself. I looked for affection and affirmation because I felt disappointed. It disappointed me that my life had not gone as planned. Medical school became a distant memory. I chose a career that would pay the bills but did not exemplify my heart. I worked in a laboratory, and I was unfulfilled. I was unable to share my need to nurture others. I felt stuck. I felt useless. I felt disappointed in the path that my life had taken.

Then it happened! One day my life changed completely! I finally became sick and tired of being disappointed. It was as if a light bulb turned on. I had longer had any reasons to be disappointed! I had no reason to be stuck! Life had not passed me by. I was still young. I still had time to complete my goals. I looked at my baby, who was now a youth and decided to do better.

Three Steps to a New You.

I had to understand that I didn't have to live as a peasant; I was royalty. Here are three tips for transitioning your life to step up in your queenship.

From Good Stock

I had to remember those that came before me. My mother was not perfect, but she was terrific. She had attempted to create an ideal life for me, so it disappointed her when I got pregnant at 14! My mother was strong, intelligent, and innovative. I had come from good stock. She was a part of the village that had raised me. The village was created by other strong women that understood working and overcoming. It was already in me to be a conqueror.

2. Remember Who You Are

I am royalty. My father is King. I had to remember that my bloodline is both spiritual and natural. I am part of a royal priesthood. I had to remind myself daily that I'm not only flesh, but I am also spirit. My faith in God was my next step in becoming new. I belong to God; He orders my steps. I have a destiny and purpose. These affirmations helped me learn to love myself. I learned how to love and forgive myself. I had to see myself as God saw me. I became renewed in my mind, which restored my vision.

3. Trust the Process

I had to let go. I had to realize that I was not in control. Once I learned to love myself, I loved everything about myself. It was not an overnight sensation. It took some time, and it continues. My past, my present, and my future are in God's hands. I had to learn to trust God with me. There were areas of my life that I was hiding from God. I wanted complete control. However, I had to give those areas to God and get healed. That's when my life changed. That's when the process of life became more manageable.

The new me was a boss indeed! I earned a second bachelor's degree in nursing, which gave me more financial independence. I applied to a master's program in nursing. I acquired a taste for entrepreneurship. I am a leader, but I am also a boss. I took sewing classes and began selling my products. I saw a niche in medical apparel, and Raiment by Rita was born.

My daughter became a homeowner in her early twenties. She graduated with her master's degree in public health and has a thriving hair business. Through her vision and innovation, she continues to enlarge her business. She is not only bright but beautiful as well. She is a model that has graced the cover of a magazine. It remains to be seen all she will do. She became a boss by watching her mom become one. She learned to juggle parenthood, business, and leisure. Her ability to maintain a balance with success is impeccable.

Although life may not take the turns that you expect, do not give up on yourself. Turn your disappointment into a springboard. You may view an unexpected pregnancy as a failure, but it's an opportunity to change your life and the life of someone else. I created my biggest cheerleader! She came here with her eyes wide open. Instead of living in disappointment, I had to show her and give her the tools to build a dynasty. She had to learn that she is also royalty!

Love Boss Lady,
Sharita Davis

A ROSE BLOOMED OUT OF THORNS

Tanesha Owens, RN (The Singing Nurse)

Enduring The Race To Victory

Who told you you should give up? Who told you that your life was over? Allow me to let you in on a little secret. Life is worth living; believe and keep going. I have been through so many disappointments in life and made many mistakes, but I kept going. Although I was raised in the church, a church family member sexually abused me as a teen. I grew up in a single-parent home and had a hard time with relationships, especially with men as a teen and through adulthood. I did not know what love was, and I tried to find love in relationships with little to no substance. Even with all of that and more, I decided not to allow the terrible experiences to hold me captive in life. I became an owner of multiple businesses, a co-author, gospel artist, registered nurse, life coach, and so much more.

It Was Good Until It Wasn't

I grew up being the oldest of four girls and one boy who was the youngest. I had a good relationship with my mom and family. I was outspoken, inquisitive, and always wanted to do things on my own. I remember us being so happy with having each other. I went to church when I was seven years old. I found a positive relationship with singing and loved to watch others sing in church. Singing and writing became how I coped, and the love I had for it was top-notch. I started singing at five. I started entering and winning talent shows, and my grandmother had me singing in the "junk-joint" to display my talents. I love singing so much; it's like I'm in another world. I am delighted, and I'm able to release it's my escape.

Growing up, I was forbidden to do many things that normal kids could do, and I couldn't understand why. I couldn't go to the movies and other places, play certain board games, and I dressed like I was someone's grandmother. I had to wear skirts and dresses all the way to the floor, scarves around my neck, even if it was 90 degrees outside, and pantyhose at school functions. I disliked church because I felt like people had animosity towards me because of my singing gift and intelligence. At the tender age of 12, I was getting dirty looks by adults, talked about around the church, and constantly judged.

To make matters worse, a close church family member sexually abused me. This abuse and mistreatment at church caused me to be emotionally disconnected, defensive, and not trust people. Although looking back, I had some good times and learned some great things, but the damage overshadowed them. It was hard to separate the church and God from the people in the church at a young age.

I couldn't wait to get old enough to make my own decisions, so I thought. Academically, I did well. I was an advanced student and took college courses in high school. While in high school, a college invited me to live on campus for a summer bridge program, and boy, was that a reality check. It was indeed a different world. They gave this little church girl who was sheltered and not really experienced in the outside world a dose of college life. The culture was so real, so vibrant that I became entrenched in it. Because the church was strict, it was my first time wearing earrings in years. My mother came to visit me at college one day, and I forgot to take the earrings off. I was called a backslider and told I wasn't saved anymore. The little bit of hope I had in the church left. I was so done with the church; I didn't care anymore.

The Fighting Temptation

I loved God and singing, and I did not want to disappoint Him, but the boys, parties, and fun were great. I felt free being around other students because they didn't judge me. I saw church couples, and I had always dreamed of being married with kids and having a life that wasn't portrayed before me. At the end of the day, I just wanted to have fun, be loved, and be accepted. I was soon to be seventeen and started hanging with associates and found myself at a teen nightclub. I had met a guy while I was dancing and he approached me. We started talking, and one thing led to another. We hooked up regularly at his house. I had feelings for him and never told him or expressed it. After a while of hooking up, I knew that our relationship-or situationship- would not last. Little did I know I had so much more to learn about people and their true intentions, especially when young and immature. It was getting wild in the world, but from my experiences and lack of godly understanding, the church was just as wild with judgment and hatred, so I lived my life as I saw fit. At least in the world, I felt accepted. So I did whatever I wanted to do when I wanted to do it.

A Life Change

I noticed I missed a cycle, smells were strange, and certain foods were disgusting. I just knew I wasn't pregnant because I was a part of a new birth control study to make sure the contraceptive wasn't affecting the bones. It must've still been in the infancy stage because, sure enough, I found out I was pregnant. I immediately was in denial. My aunt was the first person I told, and then I told my soon-to-be son's father-yeah, the guy I met on the dance floor at the nightclub. He was in shock. I could tell he was nervous when he asked me what I was going to do. I was not sure what I was going to do. I was a straight-A student in high school and was attending a magnet school for healthcare. I always had dreams of becoming a doctor. My mom taught me medical terminology at a young age, so I thought that my life was over. I had just started with a modeling agency called Barbizon. So many thoughts ran through my mind.

I thought about if I should keep the baby. I had even scraped up the money to have an abortion. I was still indecisive about getting the abortion because I felt this pregnancy could have impeded my progress. Eventually, I had to let the cat out of the bag. I was still attending church causally, and I received a call from an Evangelist who prophesied that I would have a beautiful, curly head boy and told me not to abort him if that's what I was thinking of doing. I was told that he was going to be a blessing to me. That was the confirmation I needed to keep my baby.

I was afraid that people at church would learn of my pregnancy and judge me, mainly because of all the backlash I received when I was 12. I thought about how everyone was going to react. I was told that I could no longer sing at church, and I was just devastated. I felt like singing was what made me belong and the reason for my existence, and it was getting stripped away. I could no longer do what I loved the most, the thing that kept me sane, which was singing. I felt ashamed. I let God down, and I knew I had let my mother down. I thought about all my mother's talks about boys and how a lady should carry herself. I could not bear the embarrassment. I also thought about how in the world I would take care of the child. This led me to feel depressed about my current situation and anxious about my future.

So much was on my mind. I cried and cried, but I did not let it stop me. I had my baby boy, and I knew I needed to be good for him. It motivated me to be the best that I could be for him. Along with other family members, I picked up my son and moved from Cleveland, Ohio, to Columbus, Ohio, to start our new life together. I finished high school with the help of my support system, my mother. I started going back to church. Things were turning around, but guess what? I searched for love again because of the emptiness I felt, and I had another child at 19. I felt so stupid because here I was again pregnant without being married and having the love I deserved but I would not let the stupidity hold me hostage. I had to keep moving with my two little ones. I later became a wife and had another child. Ultimately, that relationship ended with a divorce, but guess what, that did not stop my progress either. I realized the love of God is what was missing all along.

Pushing To A Breakthrough

During the chaos and heartbreak, I bossed up and graduated nursing school and began my nursing career that later led to nursing management. I was a mother of three small children trying to make it happen. I had to learn how to love myself, my children, and my close family. I had to understand that I could not allow my mistakes to define me. I built a genuine relationship with God. Pushing through amid adversity became my motto. I became a gospel recording artist and traveled, blessing everyone that I could with my gift of singing while also being a nurse, which later birthed" The Singing Nurse." I later married the man of my dreams, became a First Lady of the ministry, and had another child, even though I was still navigating through the healing process of my trauma. We are still married, and let me say this, it has not been easy, but it has all been worth it. In this marriage, I learned a lot about myself, which caused me to seek the needed help, especially if I wanted our relationship to last. I discovered we are all learning this thing called life, and giving up should never be an option. Because of past trauma, it was easy for me to disconnect emotionally and eventually run. I realized that forgiving yourself and others is essential for growth and, most importantly, liberates you.

Tools for Forward Progress

The grace and mercy of God with determination have been key contributors to my success. I decided that nothing would stop me, and God so graciously granted me the favor to achieve my goals. I was able to become successful and not perfect. Here are some steps I took to bloom out of thorns.

1. **Receiving salvation-** I gave my life to Christ and was filled with His spirit. I started a daily devotion with God that comprised worship, reading the Holy Bible, and prayer. It helped me to stay grounded and positive. I was able to build trust with the one who controls and created everything.

2. **Validate yourself-** I learned how to validate myself and cope without worrying about the opinions of others. To be happy, I learned to love myself. We should give nothing or nobody the power to alter your state of being or how you feel about yourself.

3. **Go to therapy.** Having therapy helped me to face things in my past that I did not know still affected me today. It started a healing process that put things into perspective. It helped to ease the anxiety even though I had still been functioning. It was one of the best decisions that I had made.

4. **Eat healthily, drink water, and exercise.** Having a healthy mind and body is so important. Food is your fuel and whatever you put in is what you will get out of it mentally and physically. It's a complicated yet straightforward task for people to complete, but it is so important. Exercising and eating healthy helped me have a clear mind, clear skin, and naturally good shape. Now I'm not saying that you cannot have cheat meals, which is not so healthy because most of us have them, but strive to be healthy and make better choices.

5. **Write the vision**. I literally wrote the vision for my life. This helped me to solidify what I wanted. I could stay focused on what I wanted to achieve, and if something was a distraction, it had to go.

6. **Work the plan**. I made goals, created a detailed plan to complete those goals, and went to work on them because I realized that faith without works is genuinely dead. It took discipline and a strong will to stay on track, but my babies and I were worth it. Knowing the end goal and the life I wanted to provide for my children kept the foot

on the gas. Understand and realize that you are strong with Christ, and whatever you put your mind to, you can achieve; get it done.

God, The Missing Link

The conclusion that I came up with was God was my missing piece. He was there for me through everything and kept me, even when I did not want to be kept. It was his love that kept me going. I always knew of God, but it was different having a relationship with him, despite going to a church building. If I had given up, so many things would not have been birthed, and I would not have been able to provide for my family as I do. Out of all that has gone on in my life, I created the Singing Nurse Service, co-authored a book, and started a ministry to help various people with housing, including people transitioning from foster care. Do not give up, no matter what. God has placed us here for a reason. Find your motivation and your purpose and strive for greatness. I am rooting for you. Build your relationship with God and get to work. Do not allow anyone, including yourself, to make you miss out on the best days of your life. Although we may have failed at some things, that does not make us a failure. Lift your head up, learn from mistakes, keep going and boss up!

Love Boss Lady,
Tanesha (The Singing Nurse)

MEET THE AUTHORS

Dr. Janell Jones

Dr. Janell Jones a licensed social worker in Ohio and practices as a licensed clinical therapist. In addition, she is the founder of Melanin Grace Publishing, LLC, a publishing company, a certified life coach, international best-selling author, motivational speaker, and book coach. Dr. Jones created *Mahogany and Beautiful, Bright Me,* a program for elementary school girls to help build their confidence and increase self-love, conflict resolution, and goal setting. Dr. Jones is the creator of two podcasts, *Girl! You Got This!* which is a five-minute inspiration podcast gear towards women. Her second podcast, *Pick My Book (recently created),* where Dr. Jones has candid conversations with authors about their books. Dr. Jones has a coaching program that helps women get clear and make money from their passions, skills, and expertise. She has been featured on ABC, CBS, NBC, FOX, Sheen Magazine, Formidable Magazine, Miz CEO Magazine, and Courageous Woman Magazine.

IG: @janelljonesempowers

Facebook: @janelljonesempowers

Dr. Cortesha Cowan

Dr. Cortesha Cowan, Founder/ CVO of Mothers Helping Mothers Inc., Dr. Cowan is a business birthing coach, a best-selling author and a contributing author, Loani Executive Global member, Ambassador Commonwealth Entrepreneur Club. Philanthropist, International Influencer an Advocate for teen moms who experience homelessness or lack of support. Ambassador Commonwealth Entrepreneur Club. As a female entrepreneur, she has worked hard to build a step-by-step system and amazing model for opening homes for teen and young mothers that are homeless and in need of support services. She has leveraged her professional and personal experience to empower and help teen mothers, young girls and women grow. Her passion and drive for seeing teen mothers affected by homelessness are what pushed her to open the door and establish safe housing at the Haven of Hope crisis and transitional house. And as also teamed up with Girls Pride Africa in South Africa to do the same system there. Dr. Cowan has been nominated and awarded for countless awards including the 2013 International Women's Day recognition award and was nominated by Kaplan University for the Community Service Alumni award in 2013,Outstanding Community Service from national Council Of Nero Women, Micro Entrepreneur of the year Award 2015 and in January 2016 awarded The Presidential Life- Time Achievement Award from 44th President Barack Obama and she now holds the title Global Ambassador for The Unstoppable Woman Of Purpose Global Movement, 2018 Miz CEO Entrepreneurial Award, 2018 Best of Columbus Award in the category of Civic and Social Organizations, 2019 Global leadership Award, 2020 She Inspires Award for Strength Love and Success, 2020 Award from the UK, "The Best humanitarian Award", also 2020 Award "Be the Architect of Change and Transformation" from Finland overseas . 2021 My Pink Stilettos Award, 2021 Sisterhood Global Award Dr. Cowan, also featured in magazines such as The Huffington Post, Courageous Woman Magazine, Miz CEO Magazine, Glambitious Magazine, Sheen Magazine, Dr. Cowan Articles in the Lima News Paper, The Columbus Dispatch, Columbus Business First, also seen on Channel 10 News, Good Day Columbus (WSYX) (ABC-affiliate), Daytime Columbus, Lima News. Dr. Cowan is a co-author of eleven books, Her Chronicles Volume 2, Standing in God's Grace Volume 2, The Miz CEO Book Volume 1, What Is a Courageous Woman, Girl Get Up and Win, Thrive Girl Thrive, Quarentena and Beyond Volume 1, No Matter What You Can Make It, The Price Of Greatness, The Visionary Author for "Against All Odds" and a solo author of her own book "Finding you Purpose Passion and Planning it out in 20days". As a speaker, Dr. Cowan also got the opportunity to speak at the 2019 Essence Festival, and now holds the title as Global Ambassador for The Unstoppable Woman of Purpose Global Movement

IG:@decorteshacowan

Facebook:@drcorteshacowan

Ty Wallace

Ty Wallace became a teenage mother at the age of 14 and again at the age of 19. Ty is now a serial entrepreneur who struggled her way from the bottom to the top. She grew up in drug-infested ghettos and had children while still a child herself. Her goal was to ensure that her girls did not have children while they were children, live in the ghetto, or go without life's essentials. She earned her bachelor's degree, had a fantastic career in healthcare as a lead CT/Radiology Tech and a relationship coach and all while raising two of the best girls ever!

IG:@tywallace37

Facebook: @tylovescoaching

Dr. Tanikka Price, ESQ.

Dr. Tanikka Price uses her training as a lawyer, her passion as a motivational speaker and the knowledge she gained during her traumatic upbringing to develop and train others in dealing with trauma and living their best lives. She opened Price Legal Services Ohio, LLC in 2008 and has devoted herself to serving her community as both an Attorney and currently as the Director of Education for The Healthcare Collaborative of Greater Columbus. As a Family Law Attorney, Tanikka helped families navigate through divorce, child custody, child support and adoption. In 2020, she joined the panel for Ohio Domestic Violence Network serving women who have experienced domestic violence. Ms. Price is a wife and mother of six, she has her Doctorate in Education Degree as well as an American Culture degree from Vassar College and a law degree from Capital University Law School. For more info., please visit tcpesq.com and tanikkaprice.com.

IG:@tanikkaprice

Facebook@tanikka c. price

Ciera' Hale

Ciera Hale became a teenage mother at the age of 13. She is now a serial entrepreneur, Nail Technician, Nonprofit founder/C.E.O. of New beginnings learning Center of Columbus Childcare Center, 1st Kids Inc Childcare Consulting services, and creator of S.H.E. is Me Non-profit. Ms. Hale believes that where's there is passion, there is potential. Ms. Hale uses her creativity with designing nails! She's the owner of Cee Meshine nails by Cee. Ms. Hale's goal is to reach at least one young lady through her story of overcoming.

IG: @ms.hale_2u

Facebook: @cierahale

Ruby Cole

Ruby Cole became a teenage mother at the age of 17. Mrs. Cole is a Credit Center Specialist for a southern bank and founder and owner of Agitation Custom Crafts. As the owner of Agitation Customer Crafts, her goal is to bring the customer's imagination to life. Mrs. Cole is a certified life coach that is passionate about empowering women ages 40 and up. She works to help them realize it is never too late to start over. Mrs. Cole enjoys helping women be successful and finding their S.H.E. - Smart Honorable Excellence. She has always been good at motivating, and she thanks God for leading her on that path.

IG: @personalgem76

Facebook:@latoyacole

Tasha M. Yarborough

Tasha M. Yarborough is dedicated to uplifting her family, friends, and community. She does this through sharing her time, energy, and accomplishments as an example that success is within reach. Tasha credits her spiritual walk as the source of her ability to have achieved so much in her life. She is a certified public accountant (CPA) in the state of Ohio. She holds two master's degrees with distinction from Keller Graduate School of Management: MAFM, CPA Emphasis and MBA, Finance. She also holds a bachelor's degree from The Ohio State University Fisher College of Business: BSBA, Finance. Mrs. Yarborough is a principal in five local businesses:

- Kuumba Creative Group, LLC – a business consultancy;
- KCG Publishing – an imprint with authors such as Henna Quill;
- Brooklyn's Bookmarks, LLC – her five-year old daughter's business;
- Yarborough Business Group, LLC – her husband's environmental and transportation business; and
- Kyros Intelligent Solutions, a cybersecurity consultancy.

Born and raised in Columbus, Ohio, Tasha resides in the area with her husband, Sháun, and is the mother of four children and two grandchildren. She enjoys dancing, innovating, roller skating, and writing poetry.

IG:@ tashaj803

Facebook@: Tasha Yarborough (Jefferson)

Darnisha "Shay" Davis

Darnisha "Shay" Davis became a teenage mother at the age of 14. Mrs. Davis is a native of Cleveland, Ohio but currently lives in Columbus, Ohio. Mrs. Davis took her tragedy, turned it into triumph, and believes that you can operate in your purpose despite your past. Mrs. Davis is a licensed cosmetologist in Columbus, Ohio, and the owner of Shay Lovely Braids, LLC. She is a generational curse breaker and teaches her children how to be entrepreneurs.

IG: @shaylovelybraids

Facebook:@ darnisha.callmeshay

Sharita Davis

Sharita Davis is a registered nurse who has been practicing for 15 years. She is also a licensed minister and worship leader. She enjoys sharing her love and knowledge of God, His Word, health, nutrition, and self-care. She attended Temple University earning her B.S. in Biochemistry. During that time, she became a proud member of Alpha Kappa Alpha Sorority, Inc. (Delta Mu Chapter). After working several years as a chemist, she enrolled in Drexel University's ACE Program, an accelerated nursing program and received her BSN. Her love of God and desire to complete His purpose and plan in her life motivates her to sound the alarm to this generation to be bold for Christ. Her proudest achievements are being mother to her beautiful three children Dashia, Damon, and Micah and two grandchildren Syenna and Genesis. She describes them as her wealth.

IG:@sharita_davis_rn

Facebook: @Sharita Davis

Tanesha "The Singing Nurse" Owens

Tanesha is a Registered Nurse and a in nursing management. Now being called "The Singing Nurse," she has blended the two ministries. God has tremendously blessed her in nursing and the music ministry. She appeared, creating the Jingle and singing on the Show titled The Six Little Mcghees that has aired on OWN TV. Tanehsa ministered on Dorinda Clark-Cole show, The Word Network (Dr. Belinda Scott Show), and sung at her church. She has also appeared on the Bobby Jones Gospel show with Isaac Simpson and DP. Columbus Praise Fest voted her Best Female Vocalist in Columbus, Ohio, and opened for Alexis Spight. She has collaborated with artists including Big Latto, formerly Mulatto, and Lisa Lopes (Left Eye's sister) ReignDrop Lopes on a Children's Book Project "Ooops I danced My Diaper Off" written by Reggie James Pace. In 2014 she was awarded "The Best Inspirational Singer in America," title for a and a song written and produced by James Fortune. She toured with Twinkie and Larry Clark for the Hallelujah Stage play. Tanesha has a single titled *It's You* and a newly released single titled Free on all digital platforms. In 2017, she won Best Music Video for *It›s You* from the National Independent Gospel Music Awards. Tanesha had several radio and movie appearances and singing engagements.

Tanesha became co-author of Daughter's of Triumph, now available and Against All Odds. She became the CEO of The Singing Nurse Service and other ministries.

Tanesha appreciates the quality of life with her husband, sons, daughters, family & friends.

IG:@ taneshathesingingnurse

@TheSingingNurseService

Facebook: @taneshathesingingnurse

@TheSingingNurseService

Website: www.thesingingnurseservice.com